Non-peptide Inhibitors of Proprotein Convertase Subtilisin Kexins (PCSKs): An Overall Review of Existing and New Data

Colloquium Series on Protein Activation and Cancer

Animal Models of Proprotein Convertases-mediated Carcinogenesis
Andres Klein-Szanto and Daniel E. Bassi
Fox Chase Cancer Center

Non-peptide Inhibitors of Proprotein Convertase Subtilisin Kexins (PCSKs): An Overall Review of Existing and New Data
Utpal Chandra De[1,2], Priyambada Mishra[1], Prasenjit Rudra Pal[2], Biswanath Dinda[2] and Ajoy Basak[1]
[1]University of Ottawa, OHRI, Canada and [2]Tripura University, India

The Role of Proprotein Convertases in Cancer Progression and Metastasis
Peter Metrakos
McGill University

Therapeutic Potential of Furin Inhibition
John Creemers
Katholieke Universiteit Leuven, Belgium

Non-peptide Inhibitors of Proprotein Convertase Subtilisin Kexins (PCSKs):
An Overall Review of Existing and New Data
Utpal Chandra De, Priyambada Mishra, Prasenjit Rudra Pal, Biswanath Dinda, and Ajoy Basak
www.morganclaypool.com

ISBN: 9781615044740 paperback

ISBN: 9781615044757 ebook

DOI: 10.4199/C00066ED1V01Y201209PAC003

A Publication in the

COLLOQUIUM SERIES ON PROTEIN ACTIVATION AND CANCER

Lecture #3

Series Editor: Majid Khatib, University of Bordeaux

Series ISSN Pending

Non-peptide Inhibitors of Proprotein Convertase Subtilisin Kexins (PCSKs): An Overall Review of Existing and New Data

Utpal Chandra De
Department of Medicine, Biochemistry, Microbiology & Immunology, University of Ottawa, Canada
Department of Chemistry, Tripura University, India

Priyambada Mishra
Department of Medicine, Biochemistry, Microbiology & Immunology, University of Ottawa, Canada

Prasenjit Rudra Pal
Department of Chemistry, Tripura University, India

Biswanath Dinda
Department of Chemistry, Tripura University, India

Ajoy Basak
Department of Medicine, Biochemistry, Microbiology & Immunology, University of Ottawa, Canada

COLLOQUIUM SERIES ON PROTEIN ACTIVATION AND CANCER #3

MORGAN & CLAYPOOL LIFE SCIENCES

ABSTRACT

The Ca^{+2}-dependent mammalian Proprotein Convertase Subtilisin Kexins (PCSKs) or Proprotein/Prohormone Convertases (PCs) are a family of endoproteases that play critical roles not only in normal development and metabolism but also in various physiological and pathological conditions. These were initiated by the proteolytic processing of large inactive proproteins into their shorter bioactive mature forms by the PCSK enzymes. These events take place in a highly selective, orchestrated, and stepwise manner. Among the various proprotein substrates of PCSK enzymes, particularly important are the precursor growth factors that include proPDGF-A, B, proIGF-1, 2 and proVEGF-C because of their strong implications in neoplasia initiation, progression, and metastasis. As a result of these findings, PCSK enzymes, particularly furin or PCSK3, became a major target for possible interventions of cancer via the use of their selective inhibitors. Significant progress has been accomplished in the development of peptide and protein-based PCSK inhibitors. However, non-peptide PCSK9 inhibitors are more preferable because of their drug-like and other characteristics. So far, a few non-peptide inhibitors of PCSK enzymes of various types of chemical structures have been described in the literature. These include (i) Carbocyclic compounds of diterpene and streptamine class. (ii) Nitrogen (N)-based heterocyclic compounds of various types and chemical structures such as (a) pyrrolidine bis piperazines, (b) Cu/Zn chelating terpyridine derivatives; (iii) Oxygen (O)-based Heterocyclic compounds of varying types of chemical structures such as (a) Flavonoids, (b) Coumarins of simple and dimeric types, (c) Quinonoids, (d) Iridoids; (iv) Aromatic compounds such as (a) Aryl guanidino and amidino derivatives, (b) Naphthyl fluorescein derivative, and (c) Phenyl Arsonic acids; and (v) C_2-symmetrical aromatic azo-compounds. When measured against a small peptidyl-MCA fluorogenic substrate, these inhibitors displayed IC_{50} values ranging from nM to μM. A number of these inhibitors exhibited significant anti-PCSK activity when tested in ex vivo or cell culture conditions. This article provides an overall review of all non-peptide PCSK inhibitors so far reported in the literature along with those we identified recently for the first time and not yet published. The potential implications of these molecules as biochemical, therapeutical, or clinical agents will also be discussed.

KEYWORDS

proprotein convertase subtilisin kexins (PCSKs), PCSK inhibitors, non-peptides, inhibition constant, inhibitor design, therapeutic agents, protein processing, peptide bond cleavage, protease assay, fluorogenic substrates, intramolecularly quenched fluorogenic peptides

Contents

CHAPTER 1

Introduction

Proprotein convertase subtilisin kexins (PCSKs) previously called proprotein or prohormone convertases (PCs) are a family of Ca^{+2}-dependent endoproteases, structurally related to bacterial subtilisin and yeast kexins [1–4]. These proteases cleave larger inactive proproteins and prohormones into their shorter biologically active mature forms. This theory of precursor cleavage commonly referred to as the "Prohormone Theory," which was first proposed in 1967 by Steiner [5] and Chretien et al. [6] almost simultaneously. Extensive work has been carried out during the past three decades to identify the proteases responsible for this event. This led to the discovery of the first member called PC1 or PC3 (now termed as PCSK1) in 1990 by Seidah et al. [7] and Smeekens and Steiner [8] again in parallel studies. Since then, 8 more members were discovered over a period of 14 years. These are now termed as PCSK2 to PCSK9 [9]. Based on extensive studies, it is now well understood that PCSK enzymes perform crucial proteolytic tasks in order to generate mature proteins that are involved in maintaining growth, metabolism, and many other normal physiological functions in the body. Thus, the maintenance of optimum level of PCSK activity is critical for health and regular function of the body. Any shift or imbalance in their actions may lead to serious disorder or disease states. Therefore, study of these proteases particularly their biochemical, structural, clinical, and physiological properties became so crucial. Out of 9 PCSK enzymes shown schematically in Figures 1A and B with their various characteristic domains and known post translational modifications, the first 7 members namely PCSK1–7 belong to the category of kexin-type that cleave protein bonds that are characterized by the presence of recognition motif Arg/Lys/His-X_n-Arg/Lys/X-Arg↓, where X = any amino acid except the sensitive Cys residue and n = 1, 3 or 5 [1–3]. The 8th member PCSK8 previously called SKI-1 or S1P is of pyrolysin-type that specifically cleave proteins with consensus sequence Tyr/Phe-X-X-Arg/Lys/His-X-Φ-Ψ/Φ↓, where Φ = alkyl side chain containing hydrophobic amino acid such as Leu/Ile and Ψ = small amino acid such as Gly/Ala [1–3]. The latest member namely PCSK9 originally called NARC-1 [10] has been shown to be of proteinase K-type whose protease activity has been demonstrated only for its own prodomain, cleaving at Val–Phe–Ala–Gln↓Ser–Ile–Pro site. Figures 2A–K show the alignment of amino acid sequences of all 9 members of human PCSKs, indicating their homology particularly within the subtilase catalytic domain. In this alignment capital letters = complete homology; small letters = partial

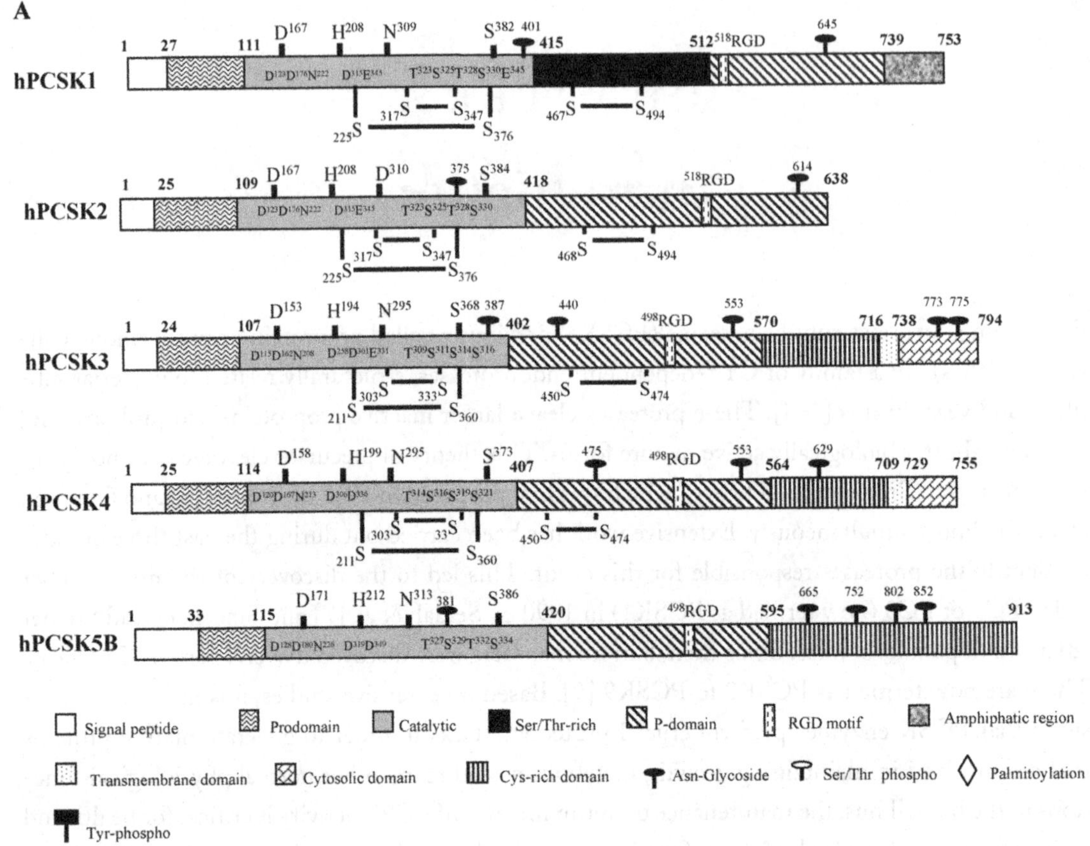

FIGURE 1A: Schematic diagram showing the structures and various characteristic domains of human (h) PCSK1 to PCSK5B.

homology and # = charged amino acid. PCSK enzymes have been implicated in various illnesses and disorders that include cancer, obesity, diabetes, hypercholesterolemia, cardiovascular, neurological diseases, viral infections, and bacterial pathogenesis [9, 11]. Of particular significance is the role of furin or PCSK3 in cancer progression, tumorigenesis and metastasis [12]. The findings were well established by various studies involving knock out and transgenic mice as well as cell transfection experiments [13–18]. As a result of this, the protease activity of PCSK enzymes particularly furin has become a major target for potential intervention of above mentioned disease and disorder conditions. Except for PCSK9, all other PCSK enzymes exhibit their functional activities via their proteolytic actions. Following this finding, tremendous interest has generated to develop inhibitors of these enzymes. A large number of compounds mostly of peptide, peptidomimetic, peptide analog, or protein origins have been reported in the literature, which exhibit varying levels of PCSK inhibitory

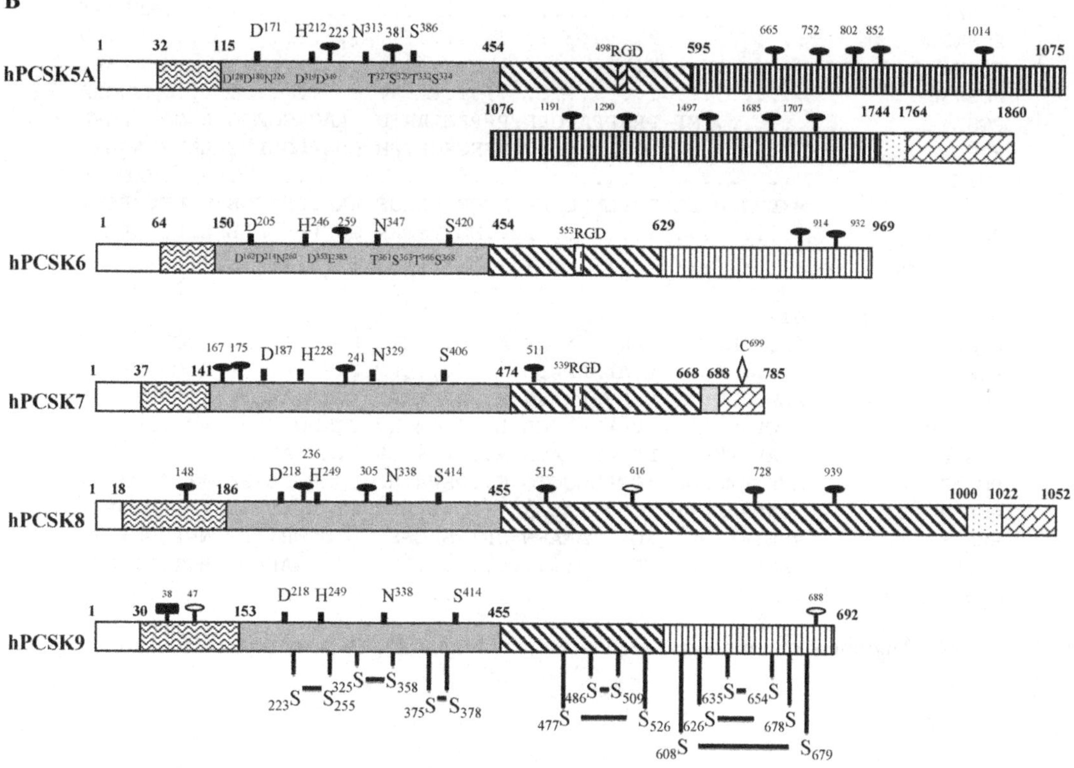

FIGURE 1B: Schematic diagram showing the structures and various characteristic domains of PCSK5A to PCSK9.

activities [19–28]. The inhibitory potency of these inhibitors range from 53.8 or 600 pM (depending on the study) [29, 30] as inhibition constant (K_i) against furin (for bioengineered α1-antitrypsin known as α1-Pdx) to relatively high μM ranges for peptide, pseudopeptide, or other derivatives [31]. Very few of them showed selectivity towards any particular member of PCSK family—an issue of great importance and challenge for their possible therapeutic applications. In addition to peptide/protein inhibitors, a limited number of small molecule non-peptide inhibitors have been reported in the literature (reviewed in [19, 20, 22, 32–34]). Owing to various reasons (explained later), non-peptide based inhibitors are always desirable and therefore increasing efforts have been devoted in recent years to develop such type of inhibitors for PCSK enzymes. A number of review articles on PCSK inhibitors and their potential therapeutic benefits and/or biochemical applications have been published in the literature during the past two decades [11, 32, 35–37]. However, these are largely devoted to protein and peptide based compounds. Moreover, since the last report more than 5 years ago, additional publications on the subject have been reported in the literature. In addition,

A

```
            1                                                   50
PCSK1       .......... .......... .......... .......... ....MERRAW
PCSK3       .......... .......... .......... .......... ....MELRPW
PCSK4       .......... .......... .......... .......... ....MRPAPI
PCSK5B      .......... .......... .......... .......... ....MGWGSR
PCSK6       ........MP PRAPPAPGPR PPPRAAAATD TAAGAGGAGG AGGAGGPGFR
PCSK7       .......... .......... MPKGRQKVPH LDAPLGLPTC LWLELAGLFL
PCSK2       .......... .......... .......... .......... ....MKGGCV
PCSK8       MKLVNIWLLL LVVLLCGKKH LGDRLEKKSF EKAPCPGCSH LTLKVEFSST
PCSK9       .......... .......... ..MGTVSSRR SWWPLPLLLL LLLLLGPAGA
Consensus   .......... .......... .......... ...p...... l.l.......

            51                                                 100
PCSK1       SLQCTAFVLF CAWCALNSAK A......KRQ FVNEWAAEI. PGGPEAASAI
PCSK3       LLWVVAATGT LVLLAADAQG .......QKV FTNTWAVRI. PGGPAVANSV
PCSK4       ALWLRLVLA. LALVRPRAVG W..APVRAPI YVSSWAVQV. SQGNREVERL
PCSK5B      CCCPGRLDLL CVLALLGGCL L..PVCRTRV YTNHWAVKI. AGGFPEANRI
PCSK6       PLAPRPWRWL LLLALPAACS A..PPPRP.V YTNHWAVQV. LGGPAEADRV
PCSK7       LVPWVMGLAG TGGPDGQGTG G..PSWAVHL ESLEGDGEE. ETLEQQADAL
PCSK2       SQWKAAAGFL FCVMVFASAE .......RPV FTNHFLVELH KGGEDKARQV
PCSK8       VVEYEYIVAF NGYFTAKARN SFISSALKSS EVDNWRIIPR NNPSSDYPSD
PCSK9       RAQEDEDGDY EELVLA.... .......LRS EEDGLAEAPE HGTTATFHRC
Consensus   .......... .......... .......... e...wa.... .g....a...
```

FIGURE 2A: Alignment of amino acid sequences of all human PCSK enzymes (PCSK1 to 9).

B

```
            101                                                150
PCSK1       AEELGYDLLG QIGSLENHYL FKHKNHPRRS RRSA...FHI TKRLSDDDRV
PCSK3       ARKHGFLNLG QIF..GDYYH FWHRGVTKRS LSPH...RPR HSRLQREPQV
PCSK4       ARKFGFVNLG PIFPDGQYFH LRHRGVVQQS LTPH...WGH RLHLKKNPKV
PCSK5B      ASKYGFINIG QIGALKDYYH FYHSRTIKRS VISS...RGT HSFISMEPKV
PCSK6       AAAHGYLNLG QIGNLEDYYH FYHSKTFKRS TLSS...RGP HTFLRMDPQV
PCSK7       AQAAGLVNAG RIGELQGHYL FVQPAGHRPA LEVEAIRQQV EAVLAGHEAV
PCSK2       AAEHGF.GVR KLPFAEGLYH FYHNGLAKAK RRRS...LHH KQQLERDPRV
PCSK8       FEVIQIKEKQ KAGLLTLEDH PNIKRVTPQR KVFRSLKYAE SDPTVPCNET
PCSK9       AKDPWRLPGT YVVVLKEETH LSQSERTARR LQAQ...AAR RGYLTKILHV
Consensus   a...g..... ..g.l...yh f......... .......... ...l.....v

            151                                                200
PCSK1       IWAEQQYEKE RSKRSAL.RD SALNLFNDPM WNQQWYLQDT RMTAALPKLD
PCSK3       QWLEQQVAKR RTKRDV.... ..YQEPTDPK FPQQWYL... ..SGVTQR.D
PCSK4       QWFQQQTLQR RVKRSV.... ..VV.PTDPW FSKQWYM... ..NSEAQP.D
PCSK5B      EWIQQQVVKK RTKRDYDFSR AQSTYFNDPK WPSMWYMHCS DNTHPCQS.D
PCSK6       KWLQQQEVKR RVKRQVR.SD PQALYFNDPI WSNMWYLHCG DKNSRCRS.E
PCSK7       RWHSEQRLLR RAKRSV.... ....HFNDPK YPQQWHLN.. ..NRRSPGRD
PCSK2       KMALQQEGFD RKKRGYRDIN EIDINMNDPL FTKQWYLINT GQADGTPGLD
PCSK8       RWSQKWQSSR PLRRASLSLG SGFWHATGRH SSRRLLRAIP RQVAQTLQAD
PCSK9       FHGLLPGFLV KMSGDLLELA LKLPHVDYIE EDSSVFAQSI PWNLERITPP
Consensus   .w...q...r r.kr....... ....h..dp. ...qw..... .........d
```

FIGURE 2B: Alignment of amino acid sequences of all human PCSK enzymes (PCSK1 to 9) (*continued*); Underlined tetrapeptide residues indicate the pro-domain cleavage sites.

C

```
            201                                                 250
PCSK1       LHVIPVWQKG ITGKGVVITV LDDGLEWNHT DIY.....AN YDPEASYDFN
PCSK3       LNVKAAWAQG YTGHGIVVSI LDDGIEKNHP DLA.....GN YDPGASFDVN
PCSK4       LSILQAWSQG LSGQGIVVSV LDDGIEKDHP DLW.....AN YDPLASYDFN
PCSK5B      MNIEGAWKRG YTGKNIVVTI LDDGIERTHP DLM.....QN YDALASCDVN
PCSK6       MNVQAAWKRG YTGKNVVVTI LDDGIERNHP DLA.....PN YDSYASYDVN
PCSK7       INVTGVWERN VTGRGVTVVV VDDGVEHTIQ DIA.....PN YSPEGSYDLN
PCSK2       LNVAEAWELG YTGKGVTIGI MDDGIDYLHP DLA.....SN YNAEASYDFS
PCSK8       V....LWQMG YTGANVRVAV FDTGLSEKHP HFKNVKERTN WTNERTLDDG
PCSK9       RYRADEYQPP DGGSLVEVYL LDTSIQSDHR EI......EG RVMVTDFENV
Consensus   ......w..g ytG.g!.!.. lDdgi...hp d........n y..e.s.#..

            251                                                 300
PCSK1       DNDHDPFPRY DPTNENKHGT RCAGEI...A MQANNHKCGV GV........
PCSK3       DQDPDPQPRY TQMNDNRHGT RCAGEV...A AVANNGVCGV GV........
PCSK4       DYDPDPQPRY TPSKENRHGT RCAGEV...A AMANNGFCGV GV........
PCSK5B      GNDLDPMPRY DASNENKHGT RCAGEV...A AAANNSHCTV GI........
PCSK6       GNDYDPSPRY DASNENKHGT RCAGEV...A ASANNSYCIV GI........
PCSK7       SNDPDPMPHP DVENGNHHGT RCAGEI...A AVPNNSFCAV GV........
PCSK2       SNDPYPYPRY TDDWFNSHGT RCAGEV...S AAANNNICGV GV........
PCSK8       LGHGTFVAGV IASMRECQGF APDAELHIFR VFTNNQVSYT SWFLDAFNYA
PCSK9       PEEDGTRFHR QASKCDSHGT HLAGVV.... ...SGRDAGV ..........
Consensus   ..d..p.p.. ..s..#.hGt rcagev.... a..nn..cgv gv........
```

FIGURE 2C: Alignment of amino acid sequences of all human PCSK enzymes (PCSK1 to 9) (*continued*); the catalytic triads D, H and S along with their catalytic hole N (or D for PCSK2) residue are shown in underlined italic fonts.

D

```
            301                                                 350
PCSK1       ......AYNS KVGGIRMLDG I.VTDAIEAS SIG.FNPGHV DIYSASWGPN
PCSK3       ......AYNA RIGGVRMLDG E.VTDAVEAR SLG.LNPNHI HIYSASWGPE
PCSK4       ......AFNA RIGGVRMLDG T.ITDVIEAQ SLS.LQPQHI HIYSASWGPE
PCSK5B      ......AFNA KIGGVRMLDG D.VTDMVEAK SVS.FNPQHV HIYSASWGPD
PCSK6       ......AYNA KIGGIRMLDG D.VTDVVEAK SLG.IRPNYI DIYSASWGPD
PCSK7       ......AYGS RIAGIRVLDG P.LTDSMEAV AFN.KHYQIN DIYSCSWGPD
PCSK2       ......AYNS KVAGIRMLDQ PFMTDIIEAS SIS.HMPQLI DIYSASWGPT
PCSK8       ILKKIDVLNL SIGGPDFMDH PFVDKVWELT ANNVIMVSAI GNDGPLYGTL
PCSK9       ......AKGA SMRSLRVLN. .....CQGKG TVSGTLIGLE FIRKSQLVQP
Consensus   ......a.n. .igg.r.$#. ...td..ea. .........i .iys.swgp.

            351                                                 400
PCSK1       DDGKTVEGPG RLAQKAFEYG VKQGRQGKGS IFVWASGNGG RQGDNCDCD.
PCSK3       DDGKTVDGPA RLAEEAFFRG VSQGRGGLGS IFVWASGNGG REHDSCNCD.
PCSK4       DDGRTVDGPG ILTREAFRRG VTKGRGGLGT LFIWASGNGG LHYDNCNCD.
PCSK5B      DDGKTVDGPA PLTRQAFENG VRMGRRGLGS VFVWASGNGG RSKDHCSCD.
PCSK6       DDGKTVDGPG RLAKQAFEYG IKKGRQGLGS IFVWASGNGG REGDYCSCD.
PCSK7       DDGKTVDGPH QLGKAALQHG VIAGRQGFGS IFVVASGNGG QHNDNCNYD.
PCSK2       DNGKTVDGPR ELTLQAMADG VNKGRGGKGS IYVWASGDGG .SYDDCNCD.
PCSK8       NNPADQMDVI GVGGIDFEDN IARFSSRGMT TWELPGGYGR MKPDIVTYGA
PCSK9       VGPLVVLLPL AGGYSRVLNA ACQRLARAGV VLVTAAGN.. FRDDACLYSP
Consensus   d.gktvdgp. .lg..af..g v..gr.g.gs ..v.asGngg ...D.c.yd.
```

FIGURE 2D: Alignment of amino acid sequences of all human PCSK enzymes (PCSK1 to 9) (*continued*); the catalytic triads D, H and S along with their catalytic hole N (or D for PCSK2) residue are shown in underlined italic fonts.

E

```
            401                                                   450
PCSK1       GYTDSIYTIS ISSASQQGLS PWYAEKCSST LATSYSSGDY TDQR..ITSA
PCSK3       GYTNSIYTLS ISSATQFGNV PWYSEACSST LATTYSSGNQ NEKQ..IVTT
PCSK4       GYTNSIHTLS VGSTTQQGRV PWYSEACAST LTTTYSSGVA TDPQ..IVTT
PCSK5B      GYTNSIYTIS ISSTAESGKK PWYLEECSST LATTYSSGES YDKK..IITT
PCSK6       GYTNSIYTIS VSSATENGYK PWYLEECAST LATTYSSGAF YERK..IVTT
PCSK7       GYANSIYTVT IGAVDEEGRM PFYAEECASM LAVTFSGGDK MLRS..IVTT
PCSK2       GYASSMWTIS INSAINDGRT ALYDESCSST LASTFSNGRK RNPEAGVATT
PCSK8       GVRGSGVKGG CRALSGTSVA SPVVAGAVTL LVSTVQKREL VNPASMKQAL
PCSK9       ASAPEVITVG ATNAQDQPVT LGTLGTNFGR CVDLFAPGED IIGASSDCST
Consensus   gy..s..t.. ...a...g.. ..y.e.c.s. la.t.s.g.. ........tt

            451                                                   500
PCSK1       DLH....NDC TETHTGTSAS APLAAGIFAL ALEANPNLTW RDMQHLVVWT
PCSK3       DLR....QKC TESHTGTSAS APLAAGIIAL TLEANKNLTW RDMQHLVVQT
PCSK4       DLH....HGC TDQHTGTSAS APLAAGMIAL ALEANPFLTW RDMQHLVVRA
PCSK5B      DLR....QRC TDNHTGTSAS APMAAGIIAL ALEANPFLTW RDVQHVIVRT
PCSK6       DLR....QRC TDGHTGTSVS APMVAGIIAL ALEANSQLTW RDVQHLLVKT
PCSK7       DWDLQKGTGC TEGHTGTSAA APLAAGMIAL MLQVRPCLTW RDVQHIIVFT
PCSK2       DLY....GNC TLRHSGTSAA APEAAGVFAL ALEANLGLTW RDMQHLTVLT
PCSK8       IASARRLPGV NMFEQGHGKL DLLRAYQILN SYKPQASLS. ...PSYIDLT
PCSK9       .........C FVSQSGTSQA AAHVAGIAAM MLSAEPELTL AELRQRLIHF
Consensus   d........c t..h.Gtsa. aplaAg.ial .l.a.p.Ltw rd.qh..v.t
```

FIGURE 2E: Alignment of amino acid sequences of all human PCSK enzymes (PCSK1 to 9) (*continued*); the catalytic triads D, H and S along with their catalytic hole N (or D for PCSK2) residue are shown in underlined italic fonts.

F

```
            501                                                   550
PCSK1       SEYDPLANN. PGWKKNGAGL MVNSRFGFGL LNAK.ALVDL ADPRTWRSVP
PCSK3       SKPAHLNAN. D.WATNGVGR KVSHSYGYGL LDAG.AMVAL A..QNWTTVA
PCSK4       SKPAHLQAE. D.WRTNGVGR QVSHHYGYGL LDAG.LLVDT A..RTWLPTQ
PCSK5B      SRAGHLNAN. D.WKTNAAGF KVSHLYGFGL MDAE.AMVME A..EKWTTVP
PCSK6       SRPAHLKAS. D.WKVNGAGH KVSHFYGFGL VDAE.ALVVE A..KKWTAVP
PCSK7       ATRYEDRRA. E.WVTNEAGF SHSHQHGFGL LNAW.RLVNA A..KIWTSVP
PCSK2       SKRNQLHDEV HQWRRNGVGL EFNHLFGYGV LDAG.AMVKM A..KDWKTVP
PCSK8       ECPYMWPYCS QPIYYGGMPT VVNVTILNGM GVTG.RIVDK PDWQPYLPQN
PCSK9       SAKDVINEAW FPEDQRVLTP NLVAALPPST HGAGWQLFCR TVWSAHSGPT
Consensus   s......... ..w..ng.g. .v.h..g.g. l.ag..lv.. a....w..v.

            551                                                   600
PCSK1       EKKECVVKDN DFEPRALKAN GEVIIEIPTR ACEGQEN.AI KSLEHVQFEA
PCSK3       PQRKCII.DI LTEPKDIGKR LEVRKT..VT ACLGEPN.HI TRLEHAQARL
PCSK4       PQRKCAV.RV QSRPTPILPL IYIREN..VS ACAGLHN.SI RSLEHVQAQL
PCSK5B      RQHVCVE.ST DRQIKTIRPN SAVRSIYKAS GCSDNPNRHV NYLEHVVVRI
PCSK6       SQHMCVA.AS DKRPRSIPLV QVLRTTALTS ACAEHSDQRV VYLEHVVVRT
PCSK7       YLASYVS.PV LKENKAIPQS PRSLEVLWNV SRMDLEMSGL KTLEHVAVTV
PCSK2       ERFHCVG.GS VQDPEKIPST GKLVLTLTTD ACEGKEN.FV RYLEHVQAVI
PCSK8       GDNIEVAFSY SSVLWPWSGY LAISISVTKK AASWEGIAQG HVMITVASPA
PCSK9       RMATAVARCA PDEELLSCSS FSRSGKRRGE RMEAQGGKLV CRAHNAFGGE
Consensus   .....!.... ..e...i... .......... a......... ..lehv....
```

FIGURE 2F: Alignment of amino acid sequences of all human PCSK enzymes (PCSK1 to 9) (*continued*).

G

```
           601                                                650
PCSK1      TIEYSRRGDL HVTLTSAAGT STVLLAERER DTSPN.GFKN WDFMSVHTWG
PCSK3      TLSYNRRGDL AIHLVSPMGT RSTLLAARPH DYSAD.GFND WAFMTTHSWD
PCSK4      TLSYSRRGDL EISLTSPMGT RSTLVAIRPL DVSTE.GYNN WVFMSTHFWD
PCSK5A     TITHPRRGDL AIYLTSPSGT RSQLLANRLF DHSME.GFKN WEFMTIHCWG
PCSK6      SISHPRRGDL QIYLVSPSGT KSQLLAKRLL DLSNE.GFTN WEFMTVHCWG
PCSK7      SITHPRRGSL ELKLFCPSGM MSLIGAPRSM DSDPN.GFND WTFSTVRCWG
PCSK2      TVNATRRGDL NINMTSPMGT KSILLSRRPR DDDSKVGFDK WPFMTTHTWG
PCSK8      ETESKNGAEQ TSTVKLPIKV KIIPTPPRSK RVLWD.QYHN LRYPPGYFPR
PCSK9      GVYAIARCCL LPQANCSVHT APPAEASMGT RVHCHQQGHV LTGCSSHWEV
Consensus  .....rrg.l ......p.gt .s...a.r.. d.....gf.. w.f...h.w.

           651                                                700
PCSK1      ENPIGTWTLR ITDMSGRIQN ...EGRIVNW KLILHGTSSQ P.........
PCSK3      EDPSGEWVLE IENTSEANNY ....GTLTKF TLVLYGTAPE ..........
PCSK4      ENPQGVWTLG LENKGYYFNT ....GTLYRY TLLLYGTAED ..........
PCSK5B     ERAAGDWVLE VYDTPSQLRN FKTPGKLKEW SLVLYGTSVQ PY........
PCSK6      EKAEGQWTLE IQDLPSQVRN PEKQGKLKEW SLILYGTAEH PYHTFSAHQS
PCSK7      ERARGTYRLV IRDVGDESFQ ...VGILRQW QLTLYGSVWS A.........
PCSK2      EDARGTWTLE LGFVGSAPQK ....GVLKEW TLMLHGTQSA PYIDQVVRDY
PCSK8      DNLRMKNDPL DWNGDHIHTN FRDMYQHLRS MGYFVEVLGA PFTCFDASQY
PCSK9      EDLGTHKPPV LRPRGQPNQ. .....CVGHR EASIHASCCH A.........
Consensus  #...g...l. ....g..... ....g.l... .l.l.g.... p.........
```

FIGURE 2G: Alignment of amino acid sequences of all human PCSK enzymes (PCSK1 to 9) (*continued*).

H

```
           701                                                750
PCSK1      .......EHM KQPRVYTSYN TVQNDRR... .......... ..........
PCSK3      ........GL PVPPE..SSG CKTLTS.... .......... ..........
PCSK4      ......MTAR PTGPQVTSSA CVQRDT.... .......... ..........
PCSK5B     .......SPT NEFPKVERFR YSRVEDPTD. .DYGTEDYAG P........C
PCSK6      RSRMLELSAP ELEPPKAALS PSQVEVPEDE EDYTAQSTPG SANILQTSVC
PCSK7      ......VDIR DRQRLLESAM SGKYLH.... .......... ..........
PCSK2      QSKLAMSKKE ELEEELDEAV ERSLKSILNK N......... ..........
PCSK8      GTLLMVDSEE EYFPEEIAKL RRDVDNGLSL VIFSDWYNTS VMRKVKFYDE
PCSK9      .......PGL ECKVKEHGIP APQ....... .......... ..........
Consensus  .......... e......... .......... .......... ..........

           751                                                800
PCSK1      ..GVEKMVDP GEEQPTQENP KENTLVSKSP SSS..SVGGR RDELEEGAPS
PCSK3      ..SQACVVCE EGFSLHQKSC VQHCPPGFAP QVL..DTHYS TENDVETIRA
PCSK4      ..EGLCQACD GPAYILGQLC LAYCPPRFFN HTR..LVTAG PGHTAAPA.L
PCSK5B     DPECSEVGCD GPGPDHCNDC LHYYYKLKNN TRI..CVSSC PPGHY.HADK
PCSK6      HPECGDKGCD GPNADQCLNC VHFSLGSVKT SRK..CVSVC PLGYFGDTAA
PCSK7      ..DDFALPCP PGLKIPEEDG YTITPNTLKT LVLVGCFTVF WTVYYMLEVY
PCSK2      .......... .......... .......... .......... ..........
PCSK8      NTRQWWMPDT GGANIPALNE LLSVWNMGFS DGLYEGEFTL ANHDMYYASG
PCSK9      ..EQVTVACE EGWTLTGCSA LPGTSHVLGA YAV.DNTCVV RSRDVSTTGS
Consensus  .......c. .g....... l......... .......... ..........
```

FIGURE 2H: Alignment of amino acid sequences of all human PCSK enzymes (PCSK1 to 9) (*continued*).

I

```
               801                                                    850
PCSK1          QAMLRLLQSA FSKNSPPKQS PKKSPSAKLN IPYENFY..E ALEKLNKPSQ
PCSK3          SVCAPCHASC ATCQGPALTD CLSCPSHASL DPVEQTCSRQ SQSSRESPPQ
PCSK4          RVCSSCHASC YTCRGGSPRD CTSCPPSSTL DQQQGSC..M GPTTPDSRPR
PCSK5B         KRCRKCAPNC ESCFGSHGDQ CMSCKYGYFL NEETNSCVTH CPDGSYQDTK
PCSK6          RRCRRCHKGC ETCSSRAATQ CLSCRRGFYH HQEMNTCVTL CPAGFYADES
PCSK7          LSQRNVASNQ VCRSGPCHWP HRSRKAKEEG TELESVPLCS SKDPDEVETE
PCSK2          .......... .......... .......... .......... ..........
PCSK8          CSIAKFPEDG VVITQTFKDQ GLEVLKQETA VVENVPILGL YQIPAEGGGR
PCSK9          TSEGAVTAVA ICCRSRHLAQ ASQELQ.... .......... ..........
Consensus      .s........ .......... .......... .......... ..........

               851                                                    900
PCSK1          L.....KDSE DSLYNDYVDV FYNTKPYKHR DDRLLQALVD ILNEEN....
PCSK3          Q.....QPPR LPPEVEAGQR LRAGLLPSHL P.EVVAGLSC AFIVL..VFV
PCSK4          L.....RAAA CPHHRCPASA MVLSLLAVTL GGPVLCGMSM DLPLY..AWL
PCSK5B         K.....NLCR KCSENCKTCT .EFHNCTECR DGLSLQGSRC SVSCEDGRYF
PCSK6          Q.....KNCL KCHPSCKKCV DEPEKCTVCK EGFSLARGSC IPDCEPGTYF
PCSK7          S.....RGPP TTSDLLAPDL LEQGDWSLSQ NKSALDCPHQ HLDVPHGKEE
PCSK2          .......... .......... .......... .......... ..........
PCSK8          IVLYGDSNCL DDSHRQKDCF WLLDALLQYT SYGVTPPSLS HSGNRQRPPS
PCSK9          .......... .......... .......... .......... ..........
Consensus      .......... .......... .......... .......... ..........
```

FIGURE 2I: Alignment of amino acid sequences of all human PCSK enzymes (PCSK1 to 9) (*continued*).

J

```
               901                                                    950
PCSK1          .......... .......... .......... .......... ..........
PCSK3          TVFLVLQLRS GFSFRGVKVY TMDRGLISYK GLPPEAWQ.. EECPSD..SE
PCSK4          SRARATPTKP QVWLPAGT.. .......... .......... ..........
PCSK5B         NGQ..DCQPC HRFCATCAGA GADGCINCTE GYFMEDGRCV QSCSISYYFD
PCSK6          DSELIRCGEC HHTCGTCVGP GREECIHCAK NFHFHDWKCV PACGEGFYPE
PCSK7          QIC....... .......... .......... .......... ..........
PCSK2          .......... .......... .......... .......... ..........
PCSK8          GAGSVTPERM EGNHLHRYSK VLEAHLGDPK PRPLPACPRL SWAKPQPLNE
PCSK9          .......... .......... .......... .......... ..........
Consensus      .......... .......... .......... .......... ..........

               951                                                   1000
PCSK1          .......... .......... .......... .......... ..........
PCSK3          EDEGRGERTA FIKDQSAL.. .......... .......... ..........
PCSK4          .......... .......... .......... .......... ..........
PCSK5B         HSSENGYKSC KKCDISCLTC NGPGFK.NCT SCPSGYLLDL GMCQMGAICK
PCSK6          EMPGLPHKVC RRCDENCLSC AGSS.R.NCS RCKTGFT... ...QLGTSCI
PCSK7          .......... .......... .......... .......... ..........
PCSK2          .......... .......... .......... .......... ..........
PCSK8          TAPSNLWKHQ KLLSIDLDKV VLPNFRSNRP QVRPLSPGES GAWDIPGGIM
PCSK9          .......... .......... .......... .......... ..........
Consensus      .......... .......... .......... .......... ..........
```

FIGURE 2J: Alignment of amino acid sequences of all human PCSK enzymes (PCSK1 to 9) (*continued*).

K

```
                1001                                                 1050
PCSK1           .......... .......... .......... .......... ..........
PCSK3           .......... .......... .......... .......... ..........
PCSK4           .......... .......... .......... .......... ..........
PCSK5B          DATEESWAEG GFCMLVKKNN LCQRKVLQQL CCKTCTFQG. ..........
PCSK6           TNHTCSNADE TFCEMVKSNR LCERKLFIQF CCRTCLLAG. ..........
PCSK7           .......... .......... .......... .......... ..........
PCSK2           .......... .......... .......... .......... ..........
PCSK8           PGRYNQEVGQ TIPVFAFLGA MVVLAFFVVQ INKAKSRPKR RKPRVKRPQL
PCSK9           .......... .......... .......... .......... ..........
Consensus       .......... .......... .......... .......... ..........
```

Accession numbers:

hPCSK1:	AAA59918
hPCSK2:	AAB32656
hPCSK3:	P09958
hPCSK4:	NP_060043
hPCSK5B:	NP_006191
hPCSK6:	AAA59998
hPCSK7:	NP_004707
hPCSK8:	Q14703
hPCSK9:	NP_777596

FIGURE 2K: Alignment of amino acid sequences of all human PCSK enzymes (PCSK1 to 9) (*continued*).

at the present time, there are no reviews in the literature that are solely devoted to non-peptide inhibitors of PCSKs whose number is gradually growing. Owing to the ever increasing interest and demand of non-peptide based PCSK inhibitors, herein we present an overview on current day information on non-peptide inhibitors of PCSK enzymes and their biochemical studies. This review also includes some of our latest data and findings on novel non-peptide PCSK inhibitors of natural origin isolated from an Indian medicinal plant with biological properties that may likely be linked to PCSK-protease activity. This review is presented in several sections as described previously in the table of contents.

• • • •

CHAPTER 2

Historical Perspectives

The significance and potential benefical applications of PCSK inhibitors in health and diseases were realized following decades of research studies including those involving knock out and transgenic animals [38, 39]. These and other studies confirmed the important functions of these proteases in normal physiological as well as pathological conditions. Today, it is well accepted that enhanced level of proteolytic activity of PCSK enzymes leading to increased proprotein maturation is one of the primary causes for the initiation, development, and/or spread of cancer, obesity, diabetes, hypercholesterolemia, neurological/cognitive dysfunctions, viral/bacterial infections, reproductive disorders, bone and osteoarthritis including defects in bone mineralization, and cartilage development [35, 36, 40]. The therapeutic benefits in targeting PCSK enzymes in a selective manner are now well recognized, and consequently modulation of PCSK activity became a major area of research. Several review articles highlighted this aspect of PCSK research. This has also generated a great deal of interest within the pharmaceutical and academic sectors for development of selective and potent PCSK inhibitors as potential therapeutic and biochemical agents. Several macromolecule and small molecule (molecular weight < 5000 Da) based PCSK inhibitors have been designed during the past two decades. There are advantages and limitations in both types of compounds. However, small molecule inhibitors are always more preferable due to their ease of accessibility via chemical means, excellent pharmacokinetic properties including increased stability to proteolysis and heat, higher degree of absorption, metabolism, bioavailability, and solubility in aqueous medium. In few cases, small molecules may be associated with high level of toxicity, poor solubility in water, and side effects. On the other hand, macromolecules such as proteins are usually unstable to proteolysis, light, and/or heat and suffer from poor cell permeability. Overall, small molecules are more attractive as drug candidates especially if they are of fully non-peptide type which is expected to provide more stability, bioavailability, and higher metabolism.

• • • •

CHAPTER 3

Class of PCSK Inhibitors

PCSK-inhibitory compounds so far reported in the literature can be classified mainly into two categories depending on their size and molecular weights. These are macromolecules (molecular weight > 5000 Da) and small molecules (molecular weight < 5000 Da). Within each category they can be of several types based on their chemical nature and structure. These are described below under separate subheadings.

3.1 MACROMOLECULES

These are largely proteins or polypeptides identified in physiological system or bioengineered through mutations of natural or wild type proteins.

3.1.1 Physiological Proteins/Polypeptides

Several protein based macromolecules have been reported to inhibit protease activity of PCSK enzymes both in vitro and ex vivo conditions. These are either found physiologically as serpins (serine proteinase inhibitors) [41] or as nonserpin proteins. These include the serpins Dspn4 from *Drosophila melanogaster* [42, 43], *Branchiostoma lanceolatum* serpin, Bl-Spn1 [44], proteinase inhibitor 8 [45], plasminogen activator inhibitor 1 (PAI-1) [46], and CRES [47]—all containing as expected a reactive site loop structure crucial for enzyme inhibitory activity [48]. Among the nonserpin PCSK inhibitory proteins, mention may be made of pro7B2 (reviewed in [49]) and proSAS [50], both of which belong to the chromogranin family.

3.1.2 Bioengineered Proteins

Proteins including serpins have been bioengineered or transformed into variants in order to develop potent and selective inhibitors of PCSK enzymes. These modified serpins contain specific mutation/s in their reactive site loops that mimic PCSK recognition motif. So far, the double mutated bioengineered α1-antitrypsin protein known as α1-Pdx that contains **R**^{355}-I-P-**R**^{358} (bold underlined amino acid residues represent the mutations) instead of wild type sequence **A**^{355}-I-P-**M**^{358} [51] has been described as the most potent inhibitor of furin (PCSK3) with K_i value 600 pM which depends on pH [52]. This inhibitor has been widely used in a variety of cell lines [52–55] and animal models [55–57] to examine its inhibitory efficacy towards furin under physiological conditions. It has been described as highly selective towards furin although some studies suggested that it also inhibits other PCSK enzymes including PCSK1, 2, 5, and 7 with much reduced potency [57]. Other bioengineered protein/serpin based PCSK inhibitors include (i) Turkey ovomucoid third domain (OMTKY3) [58], (ii) α2-Macroglobulin [59], and (iii) eglin C [60].

3.2 SMALL MOLECULES

These can be of various types such as short peptides, polypeptides, peptide analogs, peptidomimetic, or non-peptide compounds. The only common feature of these molecules is their molecular weights which are usually lower than 5,000 Da.

3.2.1 Peptides (short)

This appears to be the largest class of small molecule PCSK inhibitors, which are obtained by synthetic means. They range from 6 amino acid long SAAS protein derived peptide [61] or hexa-Arginine (all dextro or leavo forms [62–64] peptides) to as long as 83 mer polypeptide of prodomain of h (human) furin [65]. These peptides inhibit PCSK activity with IC_{50} ranging from low nM to low μM depending on the nature of PCSK enzyme and the peptide. These peptides have been tested in vitro as well as ex vivo conditions and found to block proprotein processing mediated by PCSK enzymes particularly furin [61–65].

3.2.2 Peptide Analogs

Several types of modified peptides have been described in the literature as potent to modest inhibitors of PCSK enzymes. These are largely carboxy (C)-terminal modified peptides such as peptidyl chloromethyl ketones, semicarbazones, oximes, amino ethyl benzenesulphonyl fluoride (AEBSF) (reviewed in [19]), and so on. In most cases, these inhibitors lack selectivity and potency. However, they served as important first generation lead compounds for further development.

3.2.3 Peptidomimetics

These are peptides that contain isosteric or pseudopeptide bonds at the cleavage site of a peptide substrate. Several types of pseudopeptide bonds such as oxymethylene [66], methylene amine [67] or our highly reactive unnatural amino acid such as Eda or <u>e</u>ne<u>d</u>iyne <u>a</u>mino acid [68], and so on have been incorporated at the cleavage site of a peptide substrate for this purpose. In most cases, the inhibition is only moderate except for Eda-based peptide, which inhibited furin with IC_{50}~40 nM [68]. The most potent inhibitor compound phenylacetyl-Arg-Val-Arg-4-amidinobenzylamide molecule inhibits furin with K_i 0.81 nM. Substitution of P3 Val by other amino acid residues led to other potent furin inhibitors with $K_i < 2$ nM, for those containing guanidinyl-Ala, Ile, Phe, or Tyr. The replacement of P2 Arg by Lys was also well accepted, whereas the incorporation of D-amino acids at various positions resulted in poor inhibitors. These compounds also inhibited other PCSK enzymes equally well except PCSK2 and PCSK7. They are also poor inhibitors of trypsin like enzymes. The use of the 4-amidinobenzylamide group provides convenient synthetic access to stable proprotein convertase inhibitors and derivatives as biochemical tools and for studies in cell culture [69].

3.2.4 Constrained Cyclic or Circular Peptides

Disulphide (S-S) bridged cyclic and end–end cyclized circular peptides with no terminal ends containing pseudopeptide moiety were found to be more potent inhibitors of PCSK8 enzyme compared with their linear counterparts [66]. This is the first report of a circular pseudopeptide as an inhibitor of a PCSK enzyme. A series of new peptidomimetic furin inhibitors was described based on decarboxylated arginine mimetics in the P1 position of substrate analog.

3.2.5 Non-peptides

This type of PCSK inhibitors is most attractive and particularly significant because of their high potential as therapeutic agents. In fact, several types of non-peptide inhibitors targeting a specified protease have found important therapeutic and/or biochemical applications [70]. A number of present day drugs belong to this category, and many are in the pipeline as in the case of Hepatitis C virus infection (http://hcvdrugs.com). In HIV alone, there are currently 8 protease inhibitors that have already been approved as drugs (http://www.thebody.com/content/art880.html). Cragg et al. [72, 73] in 2009 reviewed all the small molecule chemical substances approved as drugs within the period 1981–2008, and a substantial number of them belong to the category of enzyme inhibitors. A detailed list of such drugs (including their target enzyme) is shown in Table 1. A significant number of them were derived from natural sources or produced synthetic means. As expected they all possess drug like properties, meaning good absorption, bioavailability, low toxicity and metabolic properties with low molecular weights (<5,000 Da). These compounds also offer useful structural diversity.

TABLE 1: List of Approved Drugs That Target Protease Activity

DRUG NAME (TRADE AND FULL CHEMICAL NAME)	CHEMICAL STRUCTURE	COMPANY NAME	TARGETING ENZYME	ASSOCIATED DISEASE
Saquinavir [103] (Fortovase, Invirase), (2S)-N-[(2S,3R)-4-[(3S)-3-(tert-butyl carbamoyl)-decahydroisoquinolin-2-yl]-3-hydroxy-1-phenylbutan-2-yl]-2-(quinolin-2ylformamido)butanediamide	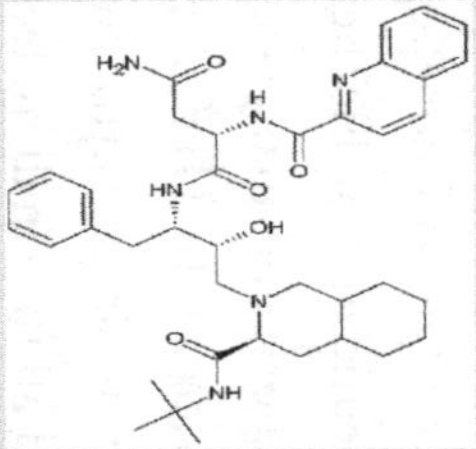	Hoffmann–La Roche	HIV-1 and HIV-2 protease	Malaria, AIDS, Cancer
Amprenavir [104] (Agenerase), (3S)-oxolan-3-yl N-[(2S,3R)-3-hydroxy-4-[N-(2-methylpropyl)(4-aminobenzene) sulfonamido]-1-phenylbutan-2-yl] carbamate	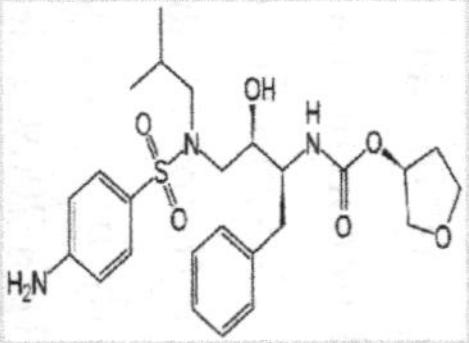	Glaxo Smith Kline	HIV protease	AIDS
Darunavir [105] (Prezista)' [(1R,5S,6R)-2,8-dioxabicyclo[3.3.0]oct-6-yl] N-[(2S,3R)-4- [(4-aminophenyl) sulfonyl- (2-methylpropyl)amino]-3-hydroxy-1-phenyl- butan-2-yl] carbamate	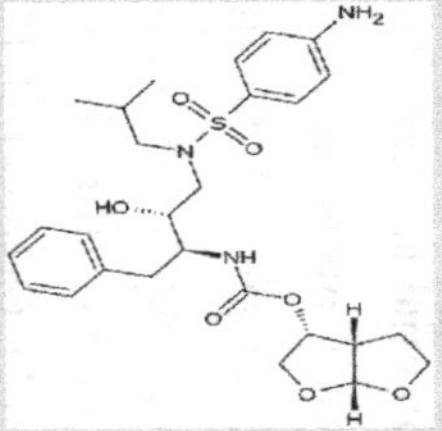	Tibotec	Reverse transcriptase	AIDS

Lopinavir [106, 107] (ABT-378), (2S)-N-[(2S,4S,5S)-5-[2-(2,6-dimethylphenoxy)acetamido]-4-hydroxy-1,6-diphenylhexan-2-yl]-3-methyl-2-(2-oxo-1,3-diazinan-1-yl) butanamide	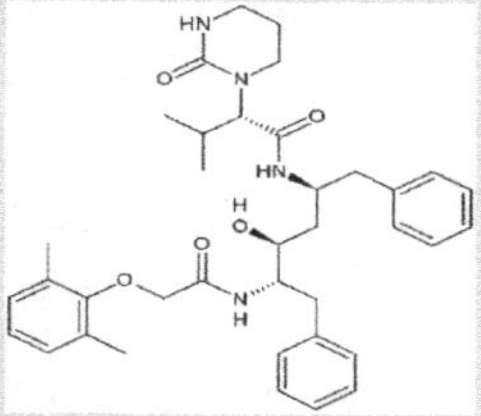	Abbott Laboratories	HIV Protease	Malaria, Retrovirus
Indinavir [108–110] (Crixivan), (2S)-1-[(2S,4R)-4-benzyl-2-hydroxy-4-{[(1S,2R)-2-hydroxy-2,3-dihydro-1H-inden-1-yl]carbamoyl}butyl]-N-tert-butyl-4-(pyridin-3-ylmethyl)piperazine-2-carboxamide	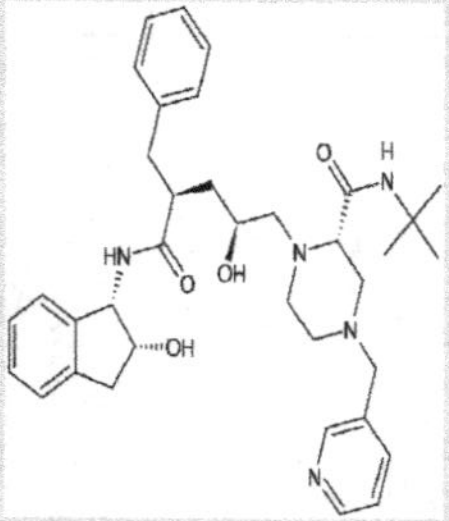	Merck & Co.	HIV protease	AIDS, Retrovirus
Ritonavir [111, 112] (Norvir), 1,3-thiazol-5-ylmethyl N-[(2S,3S,5S)-3-hydroxy-5-[(2S)-3-methyl-2-{[methyl({[2-(propan-2-yl)-1,3-thiazol-4-yl]methyl})carbamoyl]amino}butanamido]-1,6-diphenylhexan-2-yl]carbamate	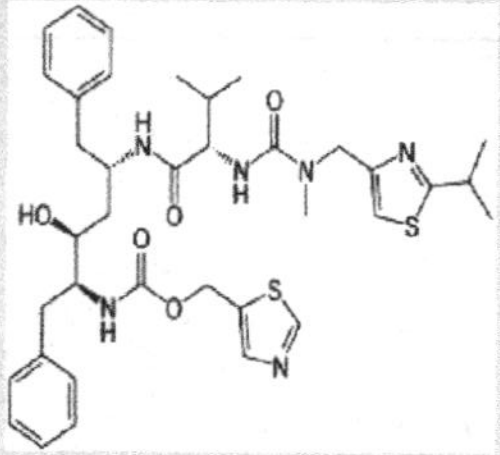	Abbott Laboratories	HIV protease	Malaria, AIDS, Cancer, Retrovirus
Fosamprenavir [113] (Lexiva, Telzir) {[(2R,3S)-1-[N-(2-methylpropyl)(4-aminobenzene)sulfonamido]-3-({[(3S)-oxolan-3-yloxy]carbonyl}amino)-4-phenylbutan-2yl]oxy}phosphonic acid	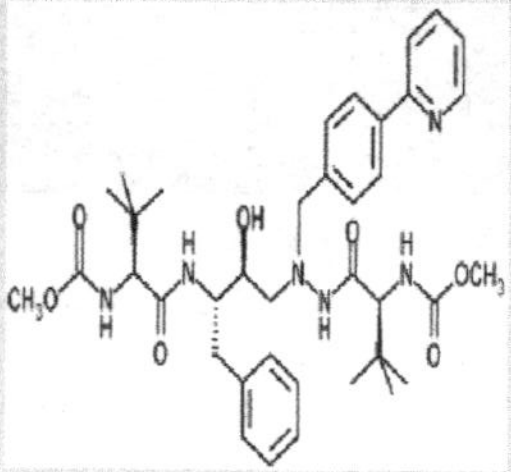	Glaxo Smith Kline	Reverse transcriptase	AIDS

TABLE 1: (*continued*)

DRUG NAME (TRADE AND FULL CHEMICAL NAME)	CHEMICAL STRUCTURE	COMPANY NAME	TARGETING ENZYME	ASSOCIATED DISEASE
Raltegravir [114] N-(2-(4-(4-fluorobenzylcarbamoyl)-5-hydroxy-1-methyl-6-oxo-1, 6-dihydropyrimidin-2-yl)propan-2-yl)	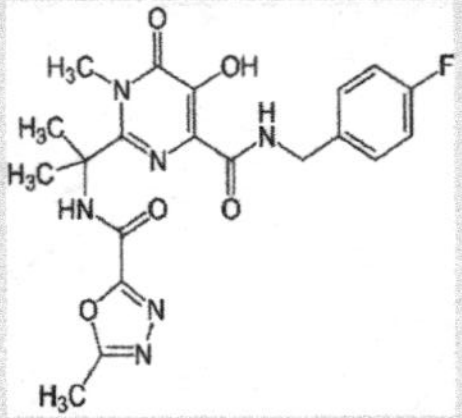	Merck & Co.	HIV integrase	AIDS
Atazanavir [115] (Reyataz) Methyl N-[(1S)-1-{[(2S,3S)-3-hydroxy-4-[(2S)-2-[(methoxycarbonyl)amino]-3,3-dimethyl-N'-{[4-(pyridin-2-yl) phenyl]methyl}butanehydrazido]-1-phenylbutan-2-yl]carbamoyl}-2,2-dimethylpropyl]carbamate	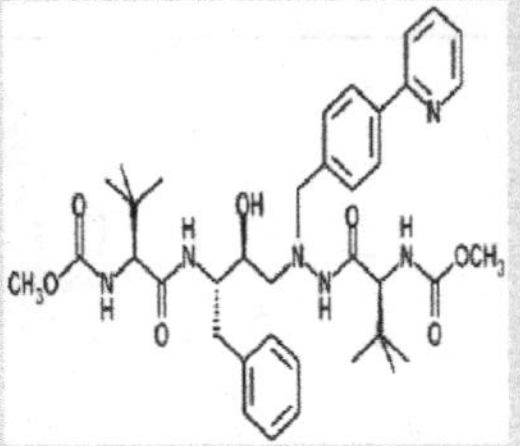	Bristol-Myers Squibb	HIV protease	AIDS
Nelfinavir [116] (Viracept) (3S,4aS,8aS)-N-tert-butyl-2-[(2R,3R)-2-hydroxy-3-[(3-hydroxy-2-methylphenyl)formamido]-4-(phenylsulfanyl) butyl]-decahydroisoquinoline-3-carboxamide	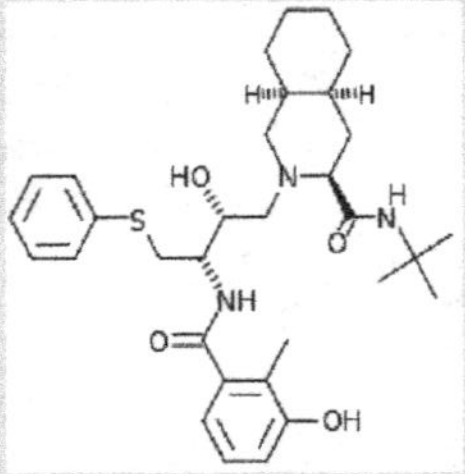	Agouron Pharmaceuticals	HIV protease	AIDS, Breast cancer

Boceprevir [117] (1R,2S,5S)-N-[(2Ξ)-4-amino-1-cyclobutyl-3,4-dioxobutan-2-yl)]- 3-{(2S)-2-[(tert-butylcarbamoyl)amino]-3,3-dimethylbutanoyl}- 6,6-dimethyl-3-azabicyclo[3.1.0]hexane-2-carboxamide	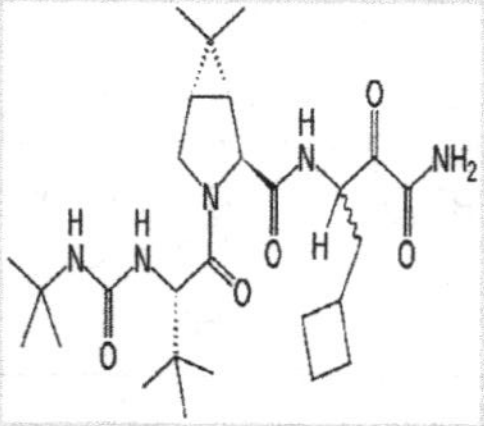	Merck & Co.	HCV nonstructural 3 (NS3) protease	Hepatitis C
Laninamivir [118, 119] (4S,5R,6R)-5-acetamido-4-carbamimidamido-6-[(1R,2R)-3-hydroxy-2-methoxypropyl]-5,6-dihydro-4H-pyran-2-carboxylic acid.	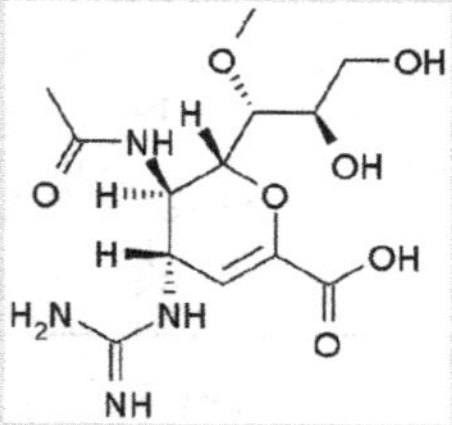	Biota Holdings Ltd.	Neuraminidase	Influenza A and B
Oseltamivir [120] (Tamiflu) Ehyl (3R,4R,5S)-5-amino-4-acetamido-3-(pentan-3-yloxy)-cyclohex-1-ene-1-carboxylate	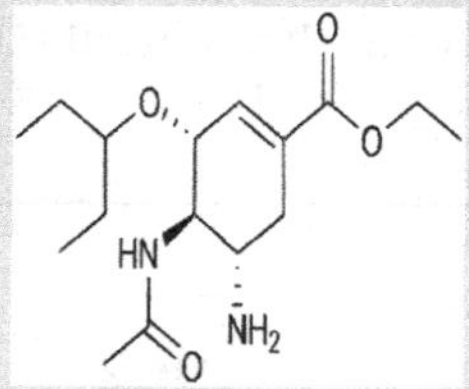	Genentech	Neuraminidase	Influenza A and B
Telaprevir [121, 122] (Incivek, Incivo) 1S,3aR,6aS)-2-[(2S)-2-[[(2S)-2-Cyclohexyl-2-(pyrazine-2-carbonylamino)acetyl]amino]-3,3-dimethylbutanoyl]-N-[(3S)-1-(cyclopropylamino)-1,2-dioxohexan-3-yl]-3,3a,4,5,6,6a-hexahydro-1H-cyclopenta[c]pyrrole-1-carboxamide	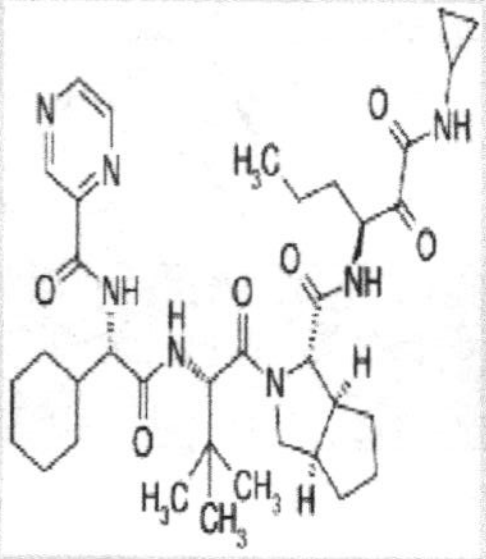	Vertex and Johnson & Johnson	Hepatitis C viral enzyme NS3.4A serine protease	Hepatitis C

TABLE 1: *(continued)*

DRUG NAME (TRADE AND FULL CHEMICAL NAME)	CHEMICAL STRUCTURE	COMPANY NAME	TARGETING ENZYME	ASSOCIATED DISEASE
Zanamivir [123, 124] (Relenza) (2R,3R,4S)-4-guanidino-3-(prop-1-en-2-ylamino)-2-((1R,2R)-1,2,3-trihydroxypropyl)-3,4-dihydro-2H-pyran-6-carboxylic acid	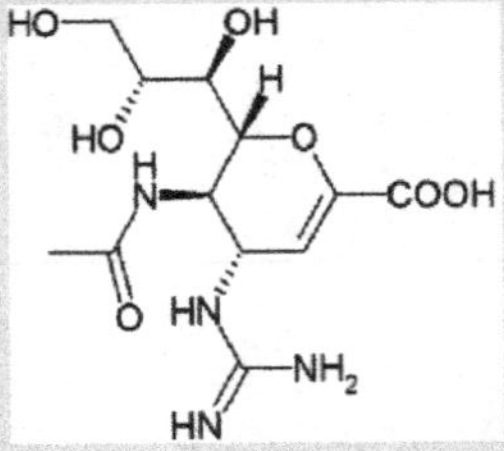	Glaxo Smith Kline	Neuraminidase	Influenza A and B
BiCNU [125] (Carmustine) 1,3-bis(2-chloroethyl)-1-nitrosourea	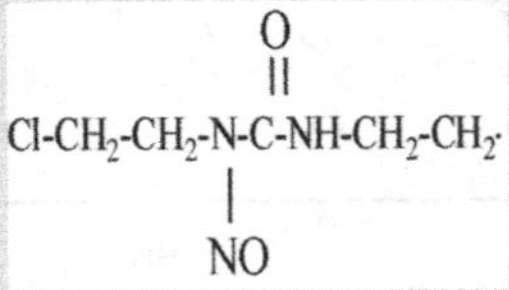 $Cl\text{-}CH_2\text{-}CH_2\text{-}N(NO)\text{-}C(=O)\text{-}NH\text{-}CH_2\text{-}CH_2\text{-}$	Bristol-Myers Squibb	Caspase	Cancer
Lamivudine [126, 127] (3TC) 4-amino-1-[(2R,5S)-2-(hydroxymethyl)-1,3-oxathiolan-5-yl]-1,2-dihydropyrimidin-2-one	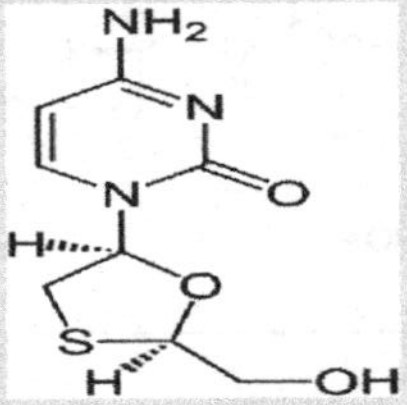	Glaxo Smith Kline	HIV reverse transcriptase	AIDS, Hepatitis B
Abacavir [128] (ABC) {(1S,4R)-4-[2-amino-6-(cyclopropylamino)-9H-purin-9-yl]cyclopent-2-en-1-yl}methanol	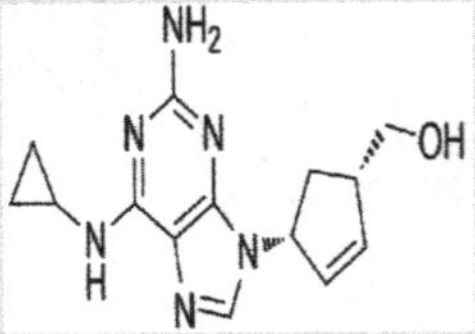	ViiV Healthcare	HIV reverse transcriptase	AIDS

Zidovudine [129, 130] (AZT) 1 [(2R,4S,5S)-4-azido-5-(hydroxymethyl) oxolan-2-yl]-5-methylpyrimidine-2,4-dione[1]	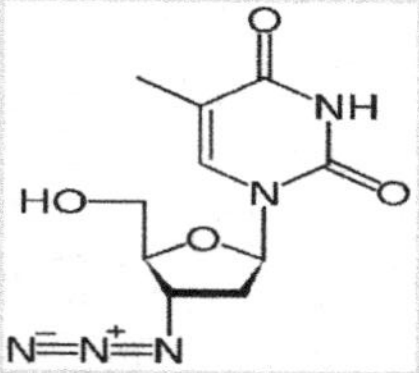	Glaxo Smith Kline	HIV reverse transcriptase	AIDS
Tipranavir [131, 132] (Aptivus) N-{3-[(1*R*)-1-[(2*R*)-6-hydroxy-4-oxo-2-(2-phenylethyl)-2-propyl-3,4-dihydro-2*H*-pyran-5-yl]propyl]phenyl}-5-(trifluoromethyl)pyridine-2-sulfonamide	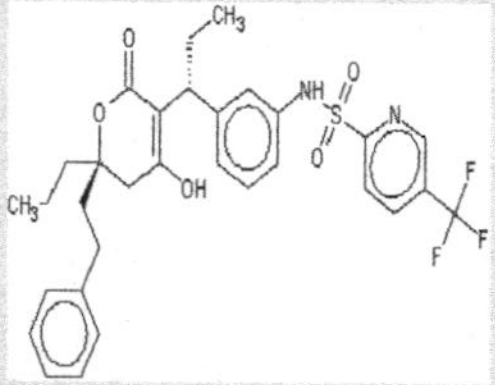	Boehringer-Ingelheim	HIV protease	AIDS

CHAPTER 4

Comparative Analyses of Peptide and Non-peptide Inhibitors

As indicated earlier, non-peptide inhibitors of PCSK enzymes are most desirable for therapeutic reason. In order to become a drug, a PCSK inhibitor must meet several criteria as laid down by Christopher Lipinski's "Rule of 5," (http://en.wikipedia.org/wiki/Lipinski's_rule_of_five) where the cut off value of each key parameter is 5 [73, 74]. Numerically, there are only four rules. It states that poor absorption or cell permeation are more likely when there are (i) more than 5 H-bond donors; (ii) the molecular weight is over 500; (iii) the Log P is over 5 meaning high lipophilicity; and (iv) the sum of N's and O's (hydrogen bond acceptor) is over 10. This empirical rule describes molecular properties important for a drug's pharmacokinetics in the human body, including their **a**bsorption, **d**istribution, **m**etabolism, and **e**xcretion ("ADME"). However, the rule does not predict if a compound is pharmacologically active. Since the above conditions are most likely to be met in most non-peptide compounds, which are in general thermally and proteolytically more stable as well as soluble in aqueous medium, interest has grown significantly to develop small molecule non-peptide inhibitors of PCSKs, which are not toxic to cells.

• • • •

CHAPTER 5

Class of Non-peptide Inhibitors of PCSKs

So far, various types of PCSK inhibitory non-peptide compounds have been described in the literature. These may be classified as follows: (i) Carbocyclic compounds; (ii) N-Heterocyclic compounds with chemical structures containing, (a) pyrrolidine bis piperazines, and (b) Cu/Zn chelating terpyridine derivatives; (iii) O-Heterocyclic compounds with varying types of chemical structures such as (a) Flavonoids, (b) Coumarins of simple and dimeric types, (c) Quinonoid, and (d) Iridoid compounds; (iv) Aromatic compounds such as (a) Aryl guanidino and amidino derivatives, (b) Naphthyl fluorescein derivative (c) Phenyl Arsonic acids, (d) Enediynes and (v) C_2-symmetrical aryl azo-compounds.

5.1 CARBOCYCLIC COMPOUNDS

5.1.1 Diterpenes

The first non-peptide inhibitor of any PCSK enzyme is andrographolide and its derivatives (compounds **1–6**) Figure 3A, which we reported in 1999. This is a naturally occurring diterpine molecule with 20 carbon atoms that belongs to labdane class [75]. It inhibits PCSK1 and furin activity with K_i in low μM values. Although found to be non-specific and modest in its action, it provided the first lead non-peptide molecule as PCSK inhibitor for future exploration.

5.1.2 Streptamines

These are originally defined as cyclohexyl diamino alcohols obtained by alkaline hydrolysis of streptomycine or streptidine and contain a common structural skeleton as shown in Figure 3B compounds **7** to **12**. Of particular interest are the 2, 5-dideoxystreptamine compounds that are derivatized with aryl guanidine function. These heterocyclic compounds have been shown by Jiao et al [76] to potently inhibit furin activity with K_i values ranging from 6–812 nM depending on the nature of the compound. Other PCSK enzymes such as PACE4 (PCSK6), PC5 (PCSK5), and PC7 (PCSK7) have also been found to be inhibited by these compounds but with moderate efficiencies. A subset of these inhibitors also exhibited antifurin activity by reducing the cleavage of anthrax toxin protein

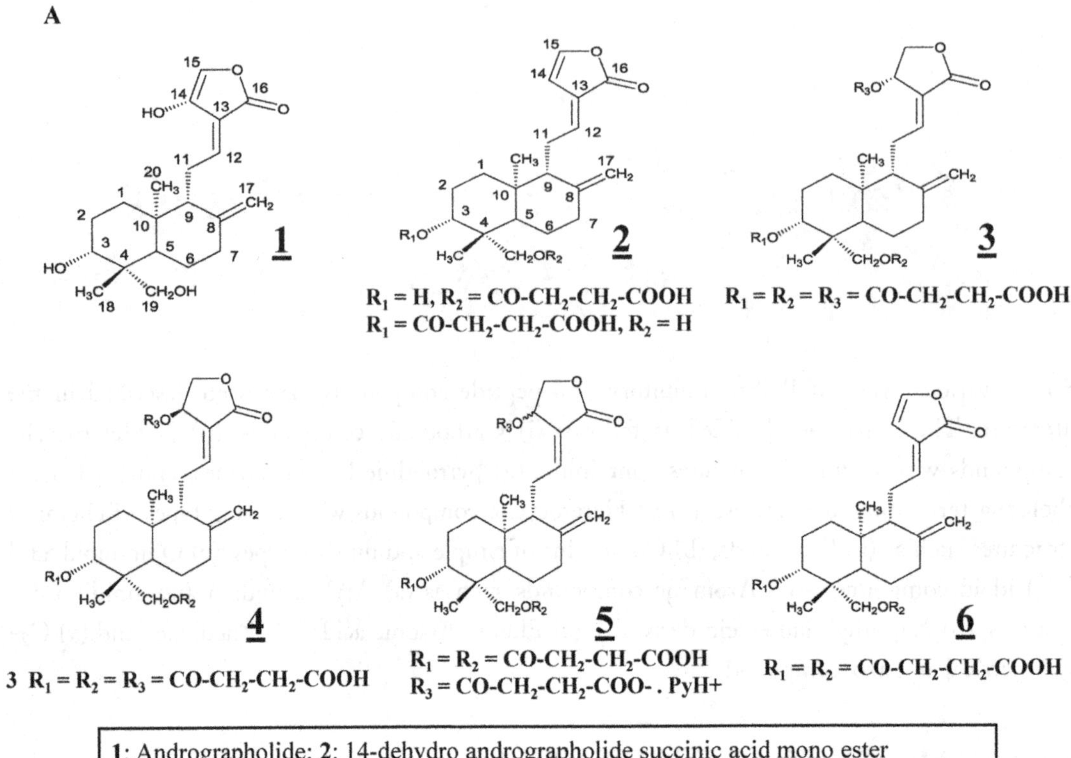

FIGURE 3A: Andrographolide derivatives displaying anti-PCSK activities in vitro.

PA by furin [77]. Furin inhibition by these compounds was found to be partly selective towards furin and competitive in nature. These inhibitors likely target the active site of furin as the molecular modeling study showed site occupancy similar to the inhibitor decanoyl-Arg-Val-Lys-Arg-CH_2Cl which alkylates furin catalytic residue.

In a recent publication, Vivoli et al [78] demonstrated that 2, 5 dideoxystreptamines upon similar modification can also inhibit PCSK1 and PCSK2 depending on the location and type of inhibition. Following screening of 45 compounds obtained by derivatization of a 2, 5-dideoxystreptamine scaffold with guanidinyl and aryl substitutions, they identified four promising PCSK1 competitive inhibitors and three PCSK2 inhibitors, which exhibited various inhibition mechanisms (competitive, non-competitive, and mixed), with sub- and low μM inhibitory potency in vitro. Certain compounds in μM concentrations blocked the processing of the physiological substrate proglucagon. The best PC2 inhibitor in this category of compounds significantly blocked

B

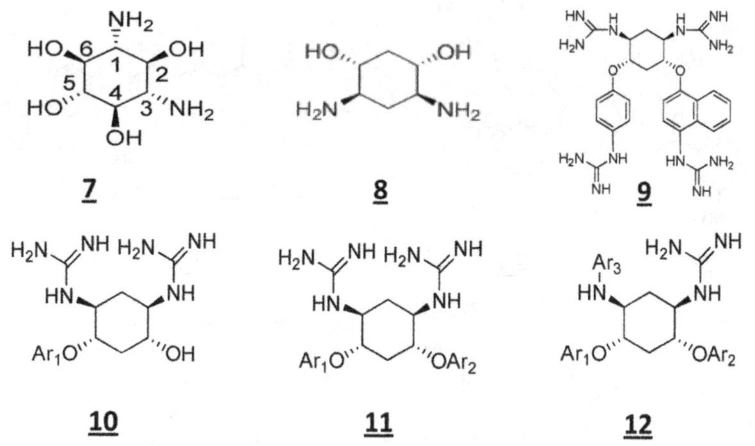

7: Streptamine	8: 2,5-Dideoxystreptamine
9: Tetra-guanidinylated diaryl 2, 5 streptamine	10: Bis-guanidinylated monoaryl-2,5-dideoxystreptamine
11: Bis-guanidinylated diaryl-2,5-dideoxystreptamine	12: Mono-guanidinylated triaryl-2,5-dideoxystreptamine

$Ar_1/Ar_2/Ar_3$ represent various aryl or aromatic rings

FIGURE 3B: Structures of various streptamine derivatives which exhibited PCSK inhibitory activities.

13: TERPY,
14: Cl-TERPY
15: TTP
16: OH-TERPY
17: MPT
18: Me_2-4'-TTP
19: tBu_3-TERPY
20: DPA
21: Me_3-[9]-aneN_3

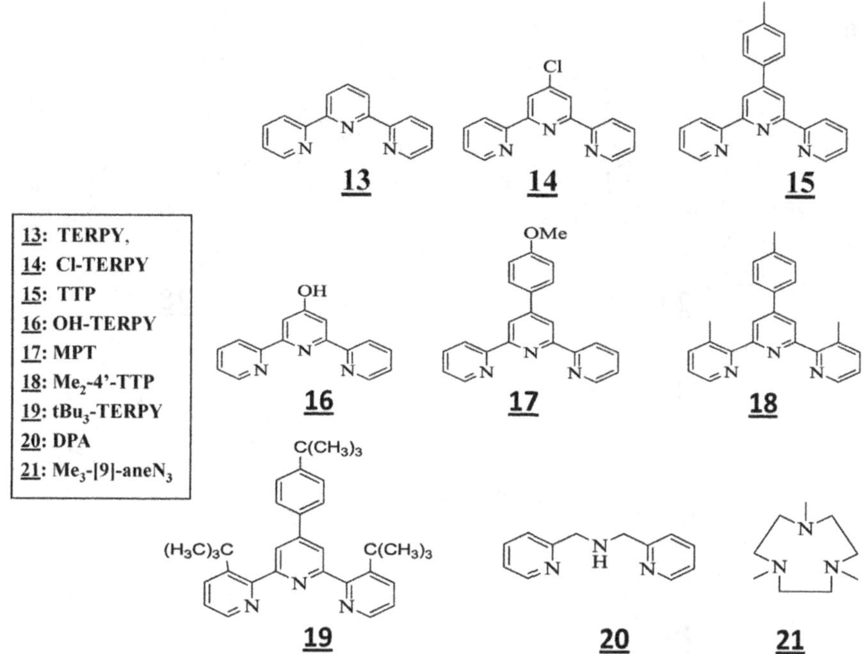

FIGURE 4: Structures of various Terpyridine derivatives exhibiting PCSK inhibitory activities.

A

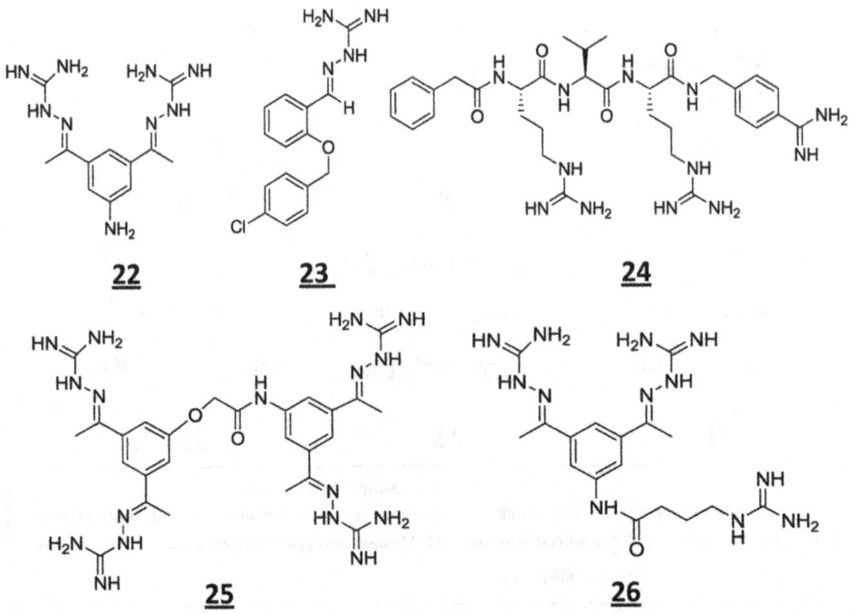

22 23 24

25 26

FIGURE 5A: Aryl amidino-hydrazone derivatives that displayed PCSK inhibitory activities.

B

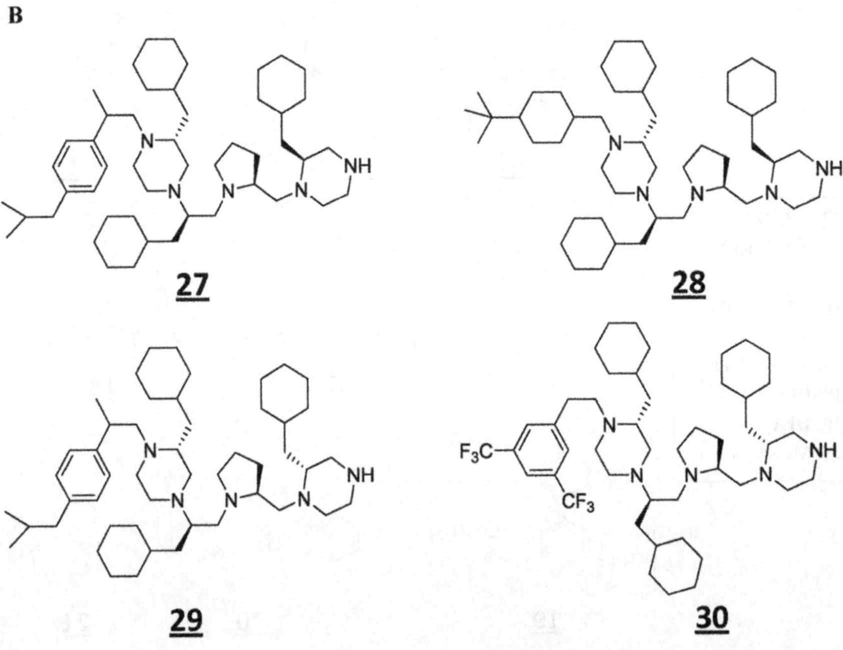

FIGURE 5B: List of pyrolidine bis-piperazines derivatives which were shown to exhibit PCSK inhibitory activities.

glucagon synthesis from its precursor proglucagon, a known PC2-mediated process, in pancreatic cell line. In addition, no cytotoxicity was observed. Interestingly, several compounds have also been identified that were able to stimulate both PCSK1 and PCSK2 activities. Such behavior has been related to the presence of aryl groups on the dideoxystreptamine scaffold. By contrast, inhibitory activity was associated with the presence of guanidinyl groups. Molecular modeling revealed specific interactions of these inhibitors with the active site of the enzyme. Some of the above compounds are proposed as bona fide lead molecules with potential therapeutic application in furin dependent diseases.

5.2 HETEROCYCLIC COMPOUNDS

In recent years, several types of heterocyclic compounds have been reported in the literature that inhibit in vitro the protease activity of various PCSK enzymes especially furin. These are generally of two types namely nitrogen and oxygen based. The detail structural features of these compounds are discussed in the sections below.

5.2.1 Nitrogen Heterocyclic Compounds

So far, two types of nitrogen based heterocyclic compounds with PCSK inhibitory activity have been reported. These are basic alkaloid type compounds and belong to terpyridine, pyrolidine, and guanidine families (Figures 4 and 5A–C) (compounds **13** to **33**).

5.2.1.1 Pyrolidines and Guanidines. Using 38 position scan libraries, Kowalska et al [79] in 2009 identified several aryl amidino hydrazones, pyrrolidine bis-piperazines, and bicyclic guanidine derivatives (Figures 5A–C) that inhibited PCSK2 activity with varying degrees of efficiencies. Even though aryl amidino/guanidino hydrazones should belong to the aromatic class, we decided to include them here because of their close similarity with the heterocyclic guanidino counterparts discussed in this section. The most potent pyrrolidine bis-piperazine displayed K_i of 0.54 μM, whereas the most potent cyclic guanidine derivative exhibited K_i of 3.3 μM against PCSK2. Cross reactivity with other PCSK enzymes were found to be limited. It was also noted that the pyrrolidine bis-piperazines exhibited $K_i > 25$ μM for PCSK1 or furin (PCSK3), whereas the K_i values of bicyclic guanidines for other convertases were found to be >15 μM. The inhibition by pyrrolidine bis-piperazines was found to be irreversible, time-dependent and non-competitive in nature. **This is the first description of small non-peptide inhibitor of PCSK2.** It was also concluded that the above types of compounds may represent promising initial leads for the optimization of therapeutically active PCSK2 inhibitors. It may be noted that PCSK2-specific inhibitors may find possible applications in the pharmacological blockade of PC2-dependent cleavage events, such as glucagon

C

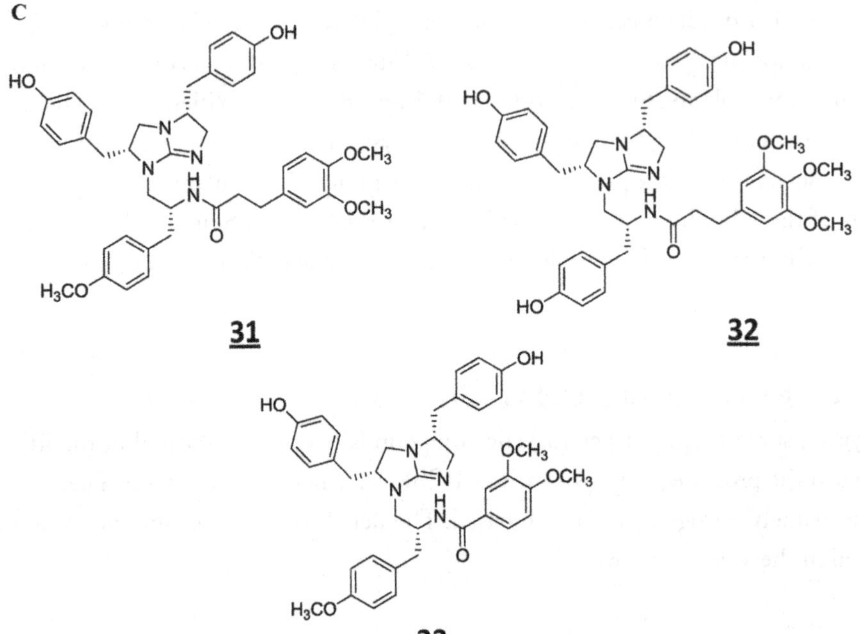

31 **32** **33**

FIGURE 5C: Bicyclic aromatic guanidine derivatives which displayed PCSK inhibitory activities.

production in the pancreas and ectopic peptide production in small-cell carcinoma as well as opioid peptides production from **P**ro-**O**pio **M**elano **C**orticoid (POMC) [80].

Using high throughput screening technique, new amino-pyrolidine amide derivatives have been identified, which selectively inhibit PCSK8 (SKI-1/S1P) activity with nm per liter efficiencies. These competitive inhibitors did not alter furin activity showing its specificity towards SKI-1 [80].

5.2.1.2 Terpyridines. In 2004, Podsiadlo et al [81] reported that small and highly stable copper and zinc complexes of the type Cu(TTP)Cl_2 and Zn(TTP)Cl_2 where TTP is 4′-[p-tolyl]-2,2′:6′,2″-terpyridine (compound **15** in Figure 4) efficiently inhibit furin (PCSK3) and the yeast counterpart Kex2 activity. The inhibition was found to be irreversible and competitive in nature. In addition, the inhibition potency was found to be affected by the nature of the substituent on the chelating molecule "TTP." It was also revealed that free chelating molecules are not inhibitors of PCSK enzymes. Also, solvated Zn^{2+} is less potent than its complexes. This is true also for copper and Kex2 enzyme. However, solvated Cu^{2+} was found to be more potent than Cu (TTP)Cl_2. A mechanism that involves coordination to the catalytic His residue was proposed for all inhibitors.

Target specificity is indicated by the fact that these metal chelating inhibitors are much less potent towards Kex2.

5.2.1.3 Heterocyclic and nonheterocyclic Enediynes. In 2009, we first reported a number of heterocyclic enediyne derivatives in particular that exhibited weak to modest inhibitory property towards furin (PCSK3) with K_i values in low to moderately high μM range [68]. Only a limited level of selectivity towards PCSK3 was observed with such compounds, and this again is depended on the nature and structure of the enediyne. Of all the compounds tested, the most potent inhibition was observed with an enediyne that is present within a 12-membered heterocyclic ring system containing a bis 1, 4 nitrogen atoms with two benzyl side chains that inhibited furin with K_i of ~10 μM [82]. However, upon incorporation of enediyne moiety into an appropriate peptide chain rather than a heterocyclic ring one can achieve an enhanced level of furin inhibition. Thus, the peptide analog QQVAKRRTKR-**Eda**-DVYQE where Eda = Enediyne amino acid inhibited furin with K_i of ~40 nM [68].

5.2.2 Oxygen Heterocyclic Compounds

Several types of oxygen heterocyclic compounds, which also contain an aromatic ring, have been identified that displayed significant PCSK inhibitory activities.

5.2.2.1 Flavonoids. In 2010, our group reported [83] PCSK inhibitory property of several naturally occurring flavonoid compounds both in vitro and ex vivo conditions. These are baicalein (compound **34**), chrysin (compound **35**), oroxylin-A (compound **36**), and oroxylin-A glucoside (compound **37**) (Figure 6), which have been isolated from the medicinal plant *Oroxylum indicum*. The measured K_i and IC_{50} values were found to be in the range of 5–35 μM. A comparative analysis of inhibition against PCSK3 (furin), PCSK4, PCSK5, and PCSK7 indicated that only baicalein exhibited a partial selectivity towards inhibition of PCSK4. A selected number of these flavonoids also blocked PCSK4-mediated processing of an intramolecularly quenched fluorogenic peptide derived from the cleavage site of its physiological substrate, human pro-insulin growth factor-1 (proIGF-1). To our understanding, this is the first demonstration of anti-PCSK activity of a flavonoid compound and also the first non-peptide inhibitor of PCSK4. Earlier, several glycosylated flavonoids such as rutin, naringin, and methyl-hesperidin were shown to inhibit furin at pH 7.2 reversibly and competitively with K_i ranging from 80 to 200 μM [84].

5.2.2.2 Coumarins: Existing and Our New Data. *(a) Dicoumarols:* Using high-throughput screens of chemical diversity libraries utilizing both enzyme-based and cell-based assays, Komiyama

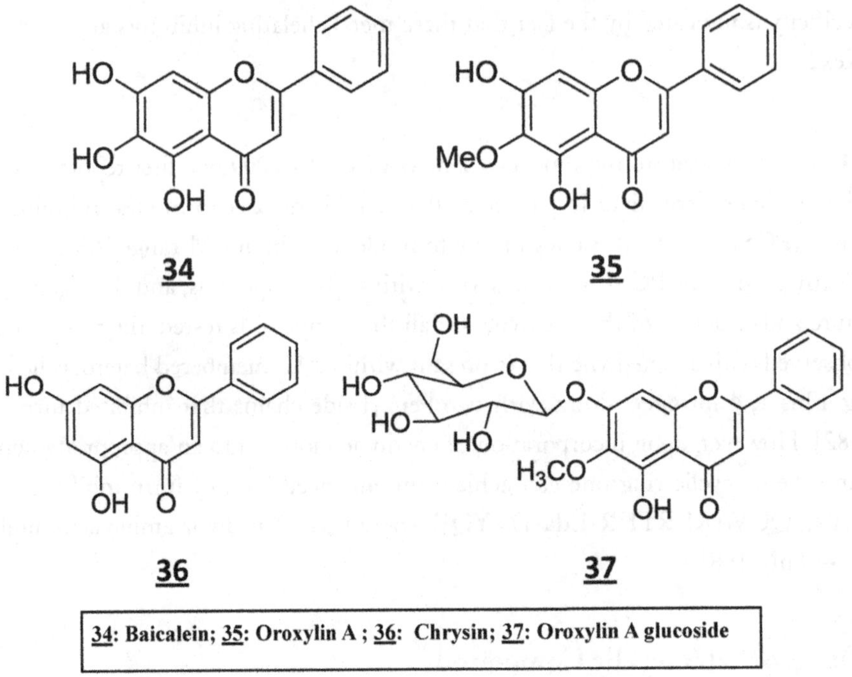

34 35

36 37

34: Baicalein; 35: Oroxylin A ; 36: Chrysin; 37: Oroxylin A glucoside

FIGURE 6: List of flavonoid compounds which exhibited PCSK inhibitory activities in vitro.

et al [85] in 2009 identified several dicoumarol derivatives structures 38 to 45 that are derived from two coumarin rings (Figure 7) as non-competitive and reversible inhibitors of furin with K_i values of 1–185 μM. These compounds inhibited furin/furin-like activity both at the cell surface and in the secretory pathway at concentrations nearly equivalent to their K_i values. However, they are not specific since other PCSK enzymes such as PCSK5, PCSK6, and PCSK7 are also inhibited with varying degrees of efficiencies. High bioavailability and relatively low toxicity of dicoumarols suggest that they can serve as good starting points for future development of drug-like inhibitors of furin and other PCSK enzymes, which can exert their activities both intracellularly and at the cell surface.

(b) Simple coumarins and our new data: Except 4-hydroxy coumarin (compound 46) and two of its derivatives (compounds 47 and 48) (Figure 8) (our data in this manuscript, see below), no other simple coumarins have been shown to inhibit PCSK activity including furin. At present, there are nearly 115 small compounds that have been identified as inhibitors of furin [86]. Out of these, 26 were analyzed by docking studies, which revealed the presence of benzimidamide moiety in the most active furin inhibitors.

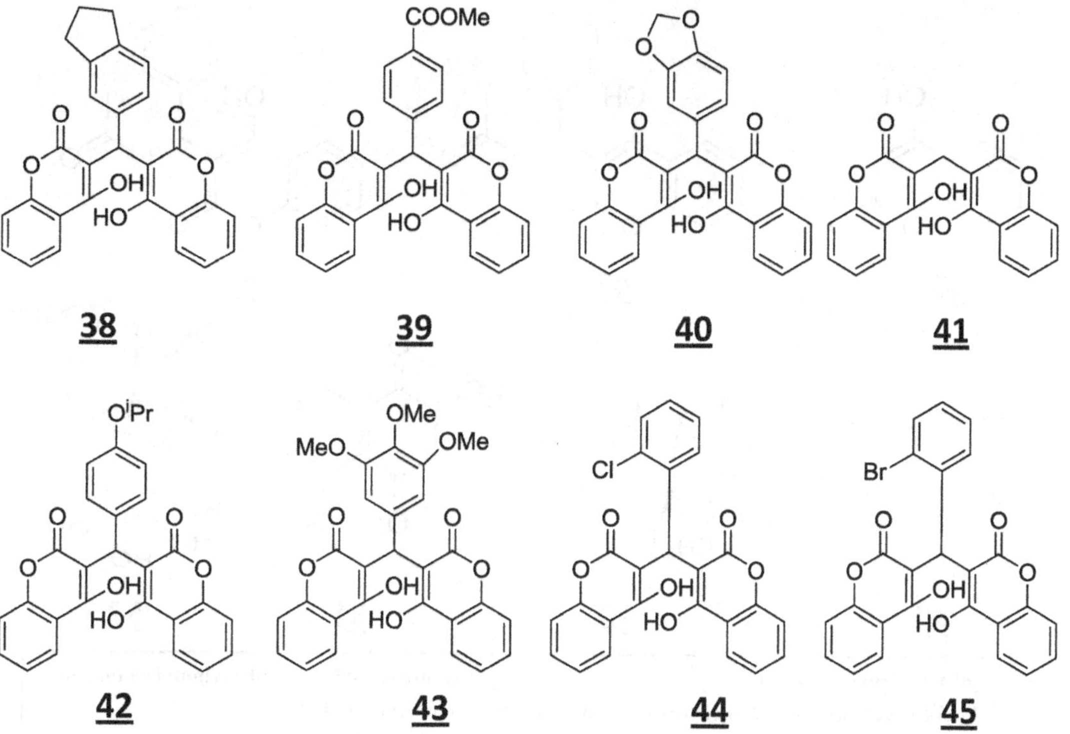

FIGURE 7: Structures of various dicoumarol derivatives exhibiting PCSK inhibitory activities.

In our own ongoing efforts to identify small molecules with PCSK inhibitory activity, we examined several Indian medicinal plants that exhibit biological activities which might be linked to protease action of PCSKs. Thus, we identified a number of simple coumarin derivatives (compounds **48** to **50**) (Figure 8) that displayed *in vitro* moderate to strong inhibition of PCSK activity especially furin. Out of three coumarins tested, one identified as "scopoletin" (referred to as MCD-1 in this manuscript as it was isolated by us from the medicinal plant *Morinda citrifolia*) stands out the most. It inhibited furin with a measured K_i of ~7 μM against the fluorogenic substrate Boc-RVRR-MCA. This was determined by both stop time and progress (on line) curve assays as previously described [82, 83] using commercially available recombinant soluble human furin (R&D Systems Inc, Minneapolis, MN USA, http://www.rndsystems.com/). Cornish–Bowden plot obtained with three different substrate concentrations was used to measure the inhibition constant K_i (Figure 9). The graph also suggested the competitive and reversible nature of inhibition. This is further confirmed by the observation that it displayed IC_{50} values (Table 2) that depend on the concentration of the

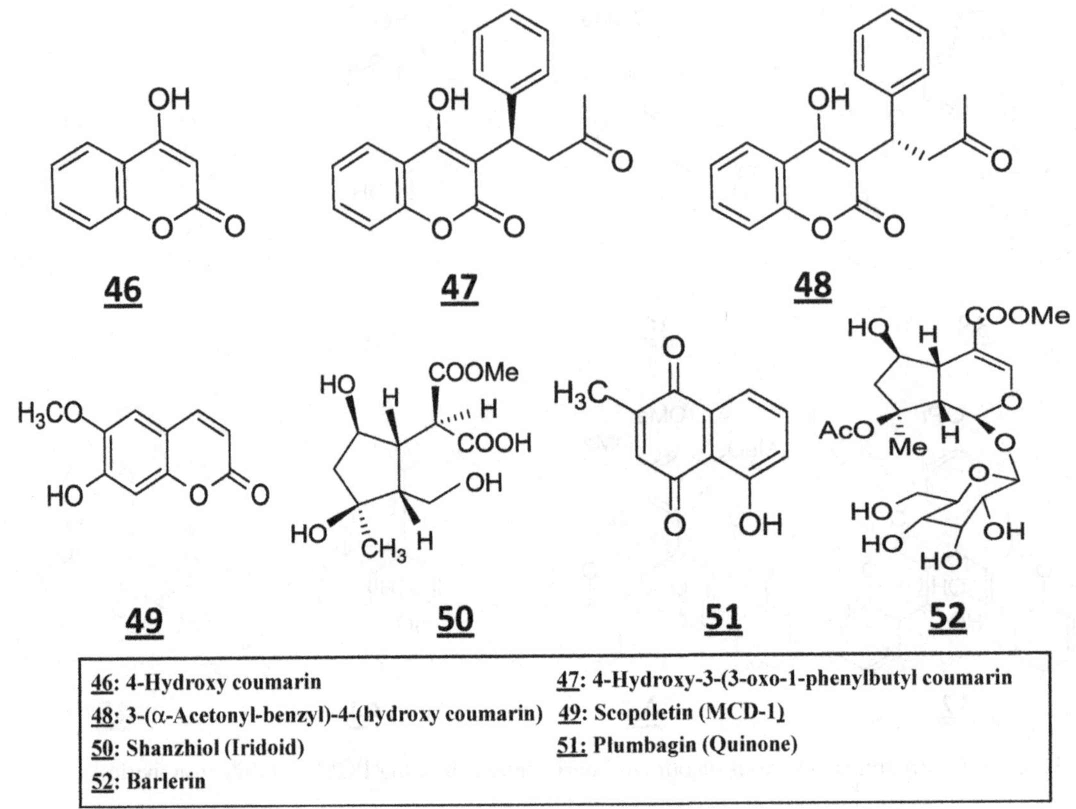

FIGURE 8: Structures of various simple coumarin, quinone and iridoid derivatives that exhibited PCSK inhibitory activities.

substrate used. All IC_{50} values were determined by using sigmoidal graph plotting the fluorescence release/hour considered as the velocity of reaction against the logarithm of concentration of the inhibitor as shown in Figure 10. On line or progress curve assay also indicated potent and gradual inhibition of furin activity in a concentration-dependent manner (Figure 11A) Using the measured IC_{50} values for various substrate concentrations and Cheng–Prusoff equation [87] (see below) valid for true competitive inhibition, we calculated the K_i value as ~4.8 μM, which was found to be in good agreement with the above measured value of 7.5 μM (see Table 2). The Michaelis–Menten constant K_m of furin used in the above equation was determined by Michaelis–Menten graph (Figure 11B). This observation provided additional support for the competitive nature of inhibition of furin by scopoletin (MCD-1).

Cheng-Prusoff equation: $K_i = IC_{50}/(1 + [S]/K_m)$, [where IC_{50} = half maximal inhibitory concentration, K_m = Michaelis–Menten constant, S = substrate concentration, K_i = inhibition constant.

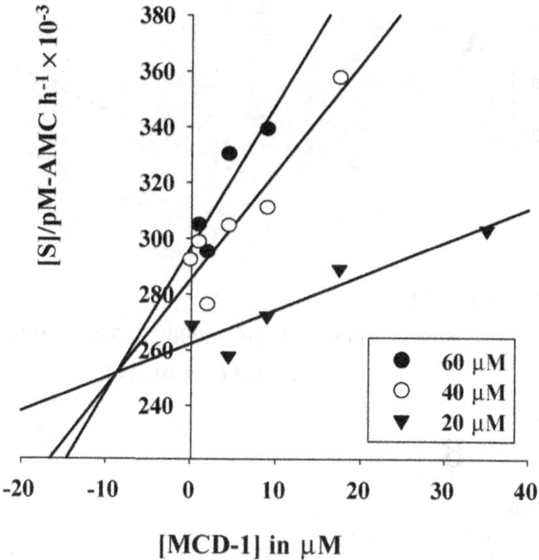

FIGURE 9: Cornish-Bowden plot showing inhibition of furin activity by MCD-1 (scopoletin) as measured with three different concentrations of substrate Boc-RVRR-MCA.

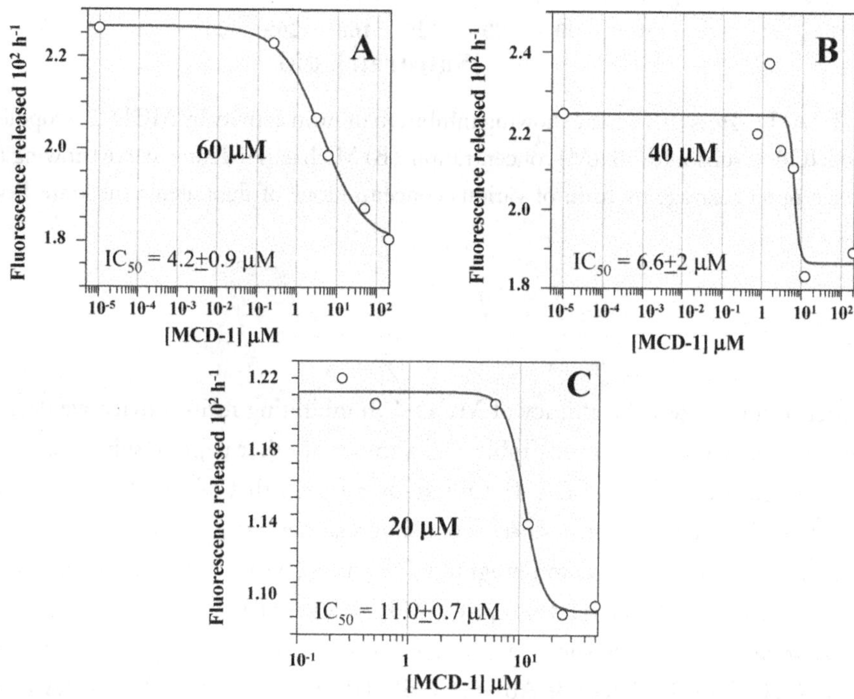

FIGURE 10: Determination of IC_{50} values for inhibition of furin activity by MCD-1 (scopoletin) using three different concentrations of substrate Boc-RVRR-MCA.

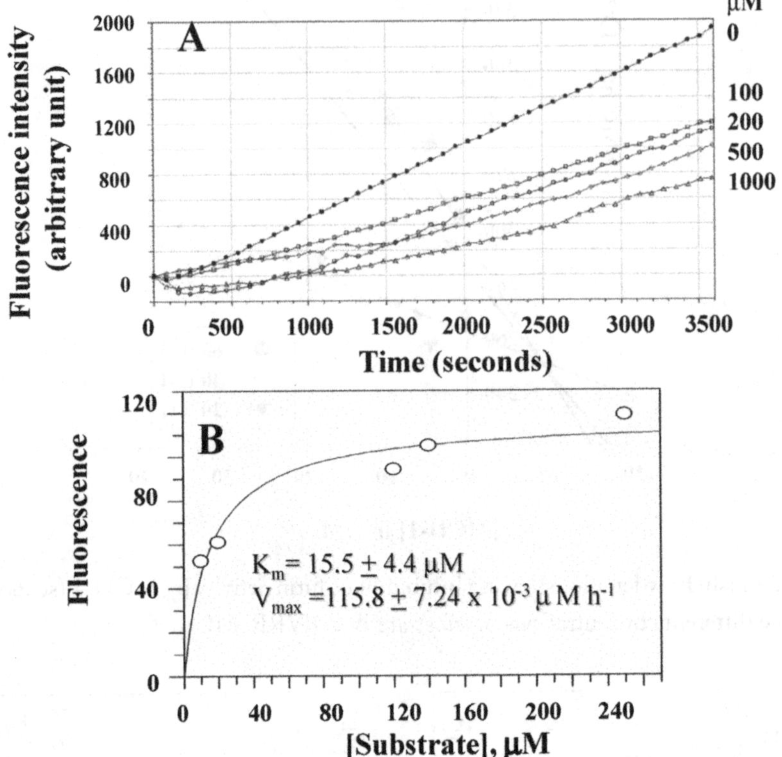

FIGURE 11: (A) Progress curve assay showing inhibition of furin activity by MCD-1 (scopoletin) using Boc-RVRR-MCA as substrate (50 μM) concentration. (B) Michaelis–Menten curve showing the release of fluorescence upon cleavage by furin of various concentrations of fluorogenic substrate Boc-RVRR-MCA.

In order to further test the efficacy of MCD-1 in inhibiting furin activity, we decided to use besides the peptidyl-MCA derivatives (Table 3) a number of other peptide substrates that contain the known cleavage site of furin (Table 4). Our study revealed that MCD-1 blocked furin mediated cleavages of these peptides in a dose-dependent manner. As a representative example, this is shown in overlaid RP-HPLC chromatograms (Figure 12A) for the fluorogenic peptide substrate Q-VEGF-C (Table 4) derived from pro-VEGF-C (Vascular Endothelial Growth factor C) furin processing site sequence as shown in Figure 13. Our data suggested that upon incubation with furin for 4h, Q-VEGF-C peptide Abz-QVHSIIRR↓SLP-Tyx-A-CO-NH_2 [Abz = 2-amino benzoic acid, Tyx = 3-nitro-tyrosine] (MW 1701, observed m/z 1698 [M^+], 1682 $[M-NH_2]^+$,

TABLE 2: Comparison of K_i values for inhibition of furin activity by MCD-1 (scopoletin) using Boc-RVRR-MCA as substrate as determined by Cornish-Bowden plot and calculated from measured IC_{50} values and K_m of furin

SUBSTRATE CONCENTRATION (μM)	IC_{50} VALUE (μM)	K_m VALUE OF FURIN (μM)	CALCULATED K_i VALUE (μM) BASED ON CHANG-PRUSOFF EQUATION	AVERAGE K_i (μM)	K_i (μM) FROM CORNISH BOWDEN PLOT
60	4.2		1.6		
40	6.6	15.5	3.7	4.8	7.5
20	11		9.1		

1665 $[M\text{-}NH_2\text{-}OH]^+$ and 1648 $[M\text{-}NH_2\text{-}2OH]^+$) (Rt = 52.2 min) is completely cleaved leading to the formation of two fragments eluting at Rt = 44.8 min (MW 1126, observed m/z 1126 $[M^+]$, 1110 $[M\text{-}NH_2]^+$, 1093 $[M\text{-}NH_2\text{-}OH]^+$) and 48.8 min (MW 593, observed m/z 593 $[M]^+$, 615 $[M+Na]^+$), respectively, which were attributed to the N-terminal (NT) [Abz-QVHSIIRR-OH] and C-terminal (CT) [SLP-Tyx-A-CO-NH_2] fragments, respectively. This suggested a cleavage at the site HSIIRR↓SLP-by furin as expected (Figure 13). The data further confirmed that MCD-1 blocked the cleavage of VEGF-C peptide in a dose-dependent manner with 50 μM MCD-1 being more potent than 5 μM as expected. We also examined possible inhibitory effect of MCD-1 on non-kexin type PCSK8 (SKI-1) enzyme. For this purpose four peptide substrates derived from Lassa virus glycoprotein, CCHFV glycoprotein site A, human CMV protease, and human SKI-1 prodomain cleavage site (Table 4) were digested with recombinant SKI-1 [66] in the absence and presence of 5 or 50 μM MCD-1. The chromatographic profile of crude digest of sample as shown for Q-CMV in Figure 12B suggested that the cleavage of this peptide by SKI-1 at the site Abz-RGVVNA↓SSRLA-Tyx-A-CO-NH_2 is efficiently blocked by 50 μM MCD-1. Other enzymes such as PCSK4, PCSK5, and PCK7 were also tested for their possible inhibition by MCD-1, but the effect was found to be minimal (Table 3). Overall our data provided the first evidence that a simple coumarin derivative such as MCD-1 or scopoletin can behave as a furin inhibitor. To our best understanding, this represents the most potent fully non-peptide inhibitor of furin.

TABLE 3: Measured IC_{50} values for inhibition of various PCSK enzymes by various coumarin, iridoid and quinone derivatives isolated from medicinal plants

COMPOUND	ENZYME	SUBSTRATE USED	IC_{50} VALUE
Scopoletin (MCD-1) (coumarin derivative)	Furin (PCSK3)	Boc-RVRR-MCA	5 μM (60 μM substrate) 8 μM (40 μM substrate) 10 μM (20 μM substrate)
	PCSK4 PCSK7	pEERTKR-MCA	>100 μM
	SKI-1 (PCSK8)	Q-SKI$^{132-142}$	
Barlerin (Iridoid glycoside derivative)	Furin (PCSK3)	Boc-RVRR-MCA	>100 μM
	PCSK4 PCSK7	pEERTKR-MCA	
	SKI-1 (PCSK8)	Q-SKI$^{132-142}$	
Shanzhiol (Iridoid derivative)	Furin (PCSK3)	Boc-RVRR-MCA	>80 μM
	PCSK4 PCSK7	pEERTKR-MCA	
	SKI-1 (PCSK8)	Q-SKI$^{132-142}$	
Plumbagin (Quinone derivative)	Furin (PCSK3)	Boc-RVRR-MCA	>100 μM
	PCSK4 PCSK7	pEERTKR-MCA	
	SKI-1 (PCSK8)	Q-SKI$^{132-142}$	

5.2.2.3 Quinonoids. In further continuation to our work, we noted for the first time that selected quinonoid compounds may also inhibit PCSK activity including furin (Table 3). This inhibitory effect especially towards furin was found to be quite modest when compared with MCD-1. Overall, the data revealed that the quinone inhibited furin *in vitro* with IC_{50} value in low μM range (Table 3).

TABLE 4: List of intramolecularly quenched fluorogenic substrates used for *in vitro* activity assay of furin and SKI-1. The table also shows the molecular weights of the peptide and its two fragments following cleavages by furin or SKI-1; Abz = 2-Amino benzoic acid, Tyx = 3-Nitro tyrosine

NAME	AMINO ACID SEQUENCE	MOLECULAR WEIGHT		
		PEPTIDE	N-TERMINAL FRAGMENT	C-TERMINAL FRAGMENT
Q-SARS	**Abz**-EQDRNTR ↓ EVFAQ-**Tyx**-CO-NH_2	1820	1038	801
Q-VEGF-C	**Abz**-QVHSIIRR ↓ SLP-**Tyx**-A-CO-NH_2	1701	1126	593
Q-GPC	**Abz**-DIYISRRLL ↓ GTFT-**Tyx**-A-CO-NH_2	1950	1266	702
Q-CMV	**Abz**-RGVVNA ↓ SSRLA-**Tyx**-A-CO-NH_2	1526	734	810
Q-CCHFV-A	**Abz**-SSGSRRLL ↓ SEES-**Tyx**-A-CO-NH_2	1706	995	729
Q-$SKI^{132-142}$	**Abz**-VFRSLK ↓ YAESD-**Tyx**-A-CO-NH_2	1709	868	860

5.2.2.4 Iridoids. While terpenes, alkaloids, coumarins, flavonoids, and quinonoids constitute major types of natural plant products, there are others organic compounds that are relatively less abundant and therefore remain less explored for their possible bioactivity especially towards proteases such as PCSKs. One of the significant members of this class of compounds is the "iridoid," [88]. So far, no study has been reported on the effect of iridoid compounds on protease activity, and we therefore became interested to examine this. Iridoids represent a large group of cyclopenta[c]pyran monoterpenoids that are found widespread in nature, mainly in dicotyledonous plant families that include *Apocynaceae*, *Scrophulariaceae*, *Diervillaceae*, *Lamiaceae*, *Loganiaceae*, and *Rubiaceae*. Recent biochemical studies revealed that iridoids exhibit a wide range of bioactivity, such as neuroprotection,

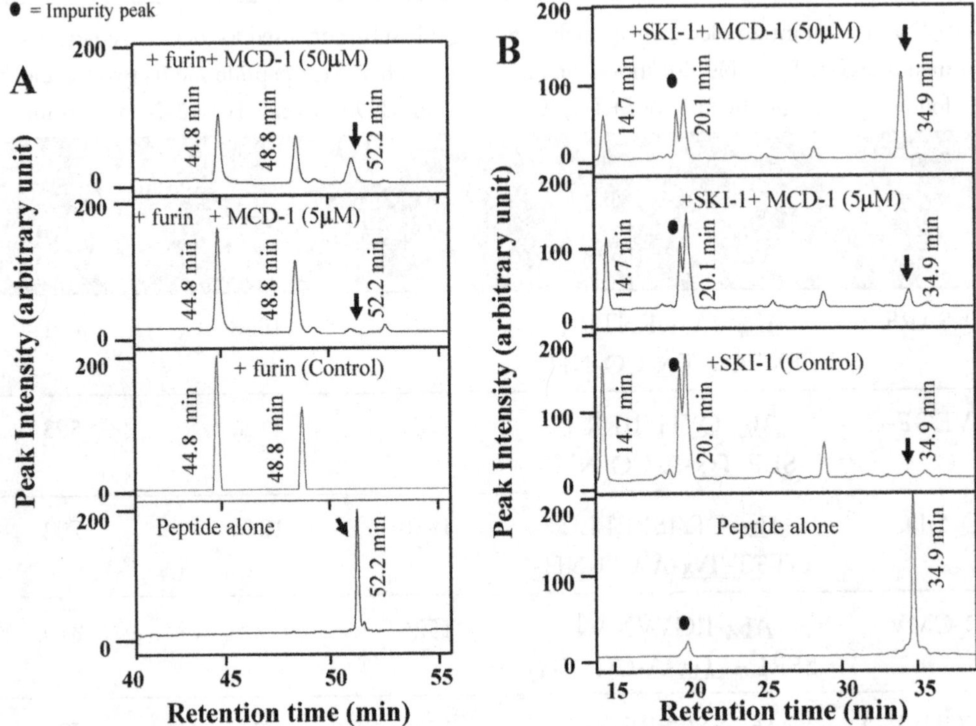

FIGURE 12: RP-HPLC chromatograms of 4h digests of Q-VEGF-C (A) and Q-CMV (B) by furin and SKI-1, respectively, in the absence & presence of various concentrations of MCD-1. * = Peak for MCD-1; • = Impurity peak.

anti-inflammatory and immunomodulatory, hepatoprotective, and cardioprotective effects. In addition antibacterial, anticancer, antioxidant, antimicrobic, hypoglycemic, hypolipidemic, choleretic, antispasmodic, and purgative properties were also reported [89]. However, so far, none of these bioactivities have ever been linked to any possible protease inhibitory property of iridoids-a notion that requires investigation. Owing to the observed role of PCSK enzymes in many of the above mentioned conditions, we examined whether iridoids possess any anti-PCSK activity or not. In fact in our study, we observed that an iridoid isolated from the plant *Mussaenda roxburghii* [90] inhibited CSK3 activity *in vitro* only with modest efficiency (Table 3). However, this finding is still interesting since it most likely represents the first demonstration of protease-inhibitory activity of any iridoid compound.

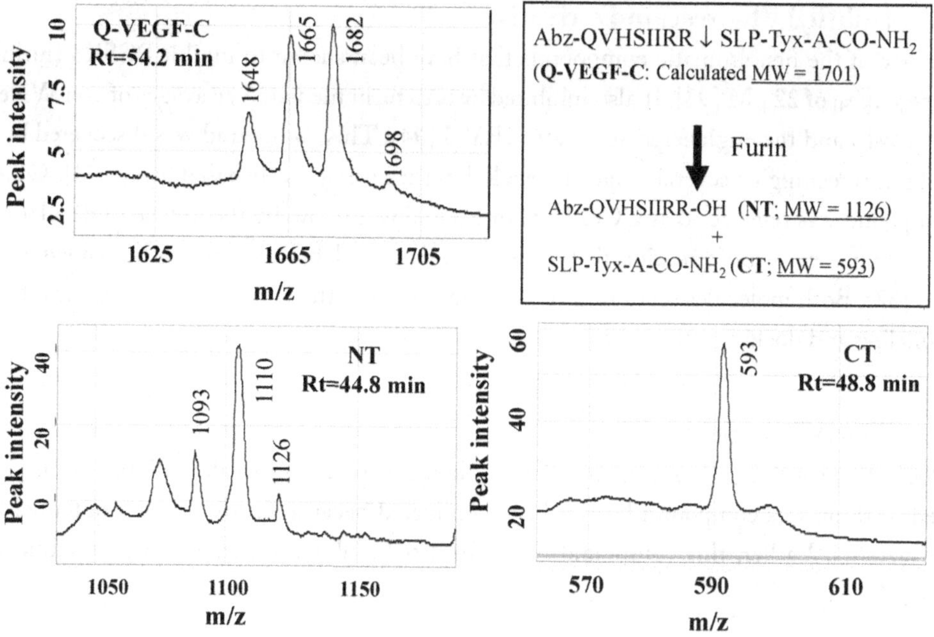

FIGURE 13: SELDI–tof mass spectra of various peaks obtained from chromatographic analysis of Q-VEGF-C following 4h digestion with furin.

5.3 AROMATIC COMPOUNDS

Several synthetic aromatic compounds particularly those containing one or multiple guanidino, amidino, or both groups have been found to inhibit PCSK3 activity. A few other types of PCSK-inhibitory aromatic compounds have also been reported. These are discussed in the sections below.

5.3.1 Simple Aryl Amidino, Guanidino, and Their Mixed Derivatives

Within the past few years, several amidino and guanidino containing aromatic hydrocarbons as well as their mixed derivatives have been synthesized, which possess moderate PCSK inhibitory activity *in vitro* against small fluorogenic peptide substrates. Of particular significance are a novel series of amidino-hydrazone-derivatives, which has been discussed already in section 5.2.1.1. However, recently simple aryl poly guanidine, poly amidine, or mixed guanidine amidino compounds have also been described as inhibitors of PCSK enzymes [91, 92].

5.3.2 Naphthyl Fluorescein Derivative

This is one of the first aromatic compounds that have been shown to inhibit PCSK3 (furin) with an average IC_{50} of 22 μM [93]. It also inhibited *ex vivo* furin mediated cleavages of von Willebrand Factor (vWF) and the engineered substrate CPA95 [94]. This compound was discovered by high-throughput screening of several small-molecule libraries using a cell-based assay [93]. Originally, the compound was referred to as CCG 8294 or B3 and was found by theoretical molecular modeling study to possess a conformation that mimics the flavonoid, baicalein—another known inhibitor of furin [83]. Both molecules seem to fit reasonably well into the catalytic pocket of furin based on 3D modeling and docking studies [93].

5.3.3 Phenyl Arsonic Acids

In a 2005 review article, I reported briefly that phenyl arsonic acid weakly inhibits furin [19]. No detailed work on this compound has yet been conducted because of toxicity and potential hazard of the material. Earlier, this compound was shown to inhibit protease activity of subtilisin and elastase [19].

5.4 C_2-SYMMETRICAL AROMATIC AZO COMPOUNDS

In 2010, we reported PCSK inhibitory property of a C_2-symmetrical compound that belonged to aromatic azo family and contains two identical parts. This compound, which possesses both E and Z-isoforms, inhibited the cellular protease PCSK8 (SKI-1/S1P) activity with K_i value ranging from 75 to 265 μM depending on the nature of the compound [95]. All inhibitions were studied with the thermally stable E-form because of the instability of the corresponding Z-form. When tested against PCSK3 (furin), these compounds did not exhibit any significant inhibition even at 100 μM concentration, suggesting their partial selectivity towards the non-kexin type PCSK enzyme, namely PCSK8, which cleaves at the carboxy terminus of a non-basic amino acid [95].

• • • •

CHAPTER 6

Comparison of Activities of Non-peptide Inhibitors of PCSK Enzymes

So far, among all the non-peptide inhibitors of PCSK enzymes identified, the coumarin, scopoletin (MCD-1) appears to be the most potent with K_i of ~7 μM against PCSK3, as reported herein for the first time. However, it is >10,000 fold less potent that the bioengineered protein α1-Pdx, which is described as the most potent PCSK3 inhibitor ever reported in the literature [29]. In terms of selectivity, not much study has been done on scopoletin. However, our recent study revealed that it also inhibited PCSK4, PCSK5, and PCSK7 to a significant extent but much less potently than for PCSK3. Most non-peptide inhibitors reported so far are significantly less potent compared to peptide or protein inhibitors with the exception of scopoletin (IC_{50} and/or K_i values in low μM range) which may be considered as a moderate inhibitor of furin. There is also a lack of specificity. Despite this limitation, they are still considered as valuable alternatives for future development of more potent and selective inhibitors of PCSK enzymes.

• • • •

CHAPTER 7

Blockade of Proprotein Processing by Non-peptide PCSK Inhibitors in Cellular Models

In contrast to peptide and protein based inhibitors of PCSK enzymes, there are only limited publications in the literature about the biochemical applications of non-peptide based PCSK inhibitors in cellular models. In this regard, one of the most studied compounds is naphthalene fluorescein derivative (B3 or CCG 8294). It was originally isolated as a bacterial side product and was shown to inhibit cancer cell motility and invasiveness [93]. Thus, it was shown that this cell-permeable compound inhibits furin-mediated cleavage of proMT1-MMP, resulting in decreased MMP-2 activation and cell motility in CHO cells expressing proMT1-MMP. As a result, it inhibited invasiveness of human fibrosarcoma cells (HT1080). It is interesting to note that this compound also inhibited other PCSKs such as PCSK6 (PACE4), PCSK5 (PC5/6), and PCSK7 (PC7), which are involved in processing substrates necessary for tumor progression and invasion [96]. The non-selectivity of this compound is actually significant since inhibition of multiple PCSKs may be necessary for complete suppression of the malignant phenotype of tumor cells.

The second type of non-peptide PCSK inhibitors tested in cellular model is the dicoumarol derivatives [85]. These compounds were found to block intracellular processing of the membrane-type 1 matrix metalloproteinase (MT1-MMP) at concentrations close to their K_i values when tested in CHO cells transfected with a plasmid encoding HA epitope-tagged MT1-MMP. Several dicoumarol compounds were also found to inhibit furin mediated processing of engineered furin substrate called CPA95 [85]. CPA95 is a secreted substrate bioengineered to contain a furin cleavage motif at residue 95. Expression of CPA95 in transfected CHO cells resulted in the appearance of both unprocessed (43 kDa) and processed (32 kDa) forms of the protein in the extracellular medium. Treatment of these cells with selected furin inhibitory dicoumarol compounds resulted in a dose-dependent decrease in the processed form and increase in the unprocessed form. These compounds also inhibited intracellular processing of pro-von Willebrand factor and were found to be mostly nontoxic to cells. In addition to the above findings, the dicoumarol compounds also block extracellular processing of anthrax protective antigen PA83 to its mature form PA63 when tested with J774A.1 murine macrophage cultures using recombinant PA83 protein. The extensive clinical use, high bioavailability, and relatively low toxicity of these compounds make them a good starting point for future development of drug-like inhibitors of PCSK enzymes that can act both intracellularly and at the cell surface.

The streptamine based furin inhibitors have also been tested in cell lines to examine their possible therapeutic benefits. Thus, it was shown that strongly furin inhibitory streptamine derivatives inhibited furin-dependent processing of anthrax PA in cultured RAW 264.7 macrophages. The PA processing was monitored by using the well-described fusion protein (FP) 59 [23, 76], in which the catalytic domain of lethal factor (LF) is replaced by that of *Pseudomonas* exotoxin A. This FP requires furin-processed PA for cellular entry and subsequent inhibition of protein synthesis, which can be quantified. It was observed that selected derivatives showed strong inhibition of PA processing with EC_{50} values ranging from 4.2 to 12.9 μM, while others inhibited PA processing to a lesser extent ($EC_{50} > 25$ μM).

PCSK2-dependent cleavage of recombinant mouse POMC (pro-opiomelano cortin) and human proglucagon has been shown to be blocked by selected pyrrolidine bis-piperazine based PCSK2 inhibitors in a dose-dependent manner. However, these effects are observed only at medium to high μM concentrations despite the fact that these compounds exhibited in vitro K_i values in low to medium nM ranges when tested against small peptide fluorogenic substrates. The reason for this observation was not explained [79].

• • • •

CHAPTER 8

Animal Study with Non-peptide PCSK Inhibitors

So far, no animal study with any non-peptide inhibitors of PCSKs has been reported in the literature although such studies will be significant in terms of their potential therapeutic benefits. In addition, *in vivo* toxicological study of these inhibitors will be needed for any further evaluation of their potential applications as biochemical, clinical and/or therapeutic agents.

• • • •

CHAPTER 9
Future Perspectives of Non-peptide PCSK Inhibitors and Concluding Remarks

Despite an ever-increasing volume of research on protein based macromolecule inhibitors of PCSKs and their useful biochemical applications, small molecule inhibitors are considered as more desirable, useful, and attractive as potential therapeutic agents. Consequently, in recent years, more attention has been devoted to develop small peptide, peptidomimetic, or fully non-peptide inhibitors of PCSKs. Moreover, a number of studies showing inhibition of intracellular processing of a variety of precursor proteins, such as growth factors, neuropeptides, receptors, viral proteins, bacterial toxins, etc, by PCSKs in the presence of their inhibitors, such as α1-Pdx, have been extensively reported in the literature [97–100]. These confirm their great potential as antitumor, antiviral, antipathogenic, and antidiuretic agents. Among the studies on small molecule PC inhibitors conducted so far, decanoyl-RVKR-chloromethyl ketone, a peptide conjugate is by far the most widely used compound. This general kexin-type PCSK inhibitor has been shown to regulate reproduction during early embryogenesis, adipocyte differentiation, and in neuronal and hormonal disorders, block viral replication, inhibit tumor cells proliferation and metastasis in cancer, prevent cartilage breakdown, and prevent proendothelin processing [101, 102]. Unlike the above two inhibitors, unfortunately up until now, no non-peptide PCSK inhibitor has made any real progress. This may primarily due to several factors that include among others lack of specificity and potency, possible toxic effects, lack of cell permeability, and insolubility in aqueous medium. These pose serious challenges for the development of non-peptide PCSK inhibitors for possible therapeutic benefits. Further research in this field including animal studies will be required.

• • • •

Acknowledgments

The authors would like to thank NSERC (Discovery), Heart & Stroke Foundation of Ontario and Center for Catalysis Research and Innovation, U Ottawa (AB), as well as Department of Science and Technology, New Delhi, India (BD) and CSIR, New Delhi, Government of India (UCD) for the funding. The above funders for this study had no role in the study design, data collection and analysis, decision to publish, or preparation of the manuscript.

References

[1] Seidah, N.G. (2011) The proprotein convertases, 20 years later. *Methods Mol Biol.* 768, pp. 23–57

[2] Steiner, D.F. (1998) The proprotein convertases. *Curr Opin Chem Biol.* 2(1), pp. 31–39.

[3] Seidah, N.G., Prat, A. (2002) Precursor convertases in the secretory pathway, cytosol and extracellular milieu. *Essays Biochem.* 38, pp. 79–94.

[4] Seidah, N.G., Chrétein, M. (1999) Proprotein and prohormone convertases: a family of subtilases generating diverse bioactive polypeptides. *Brain Res.* 848, pp. 45–62.

[5] Steiner, D.F. (1967) Evidence for a precursor in the biosynthesis of insulin. *Trans N Y Acad Sci.* 30(1), pp. 60–8.

[6] Chrétien, M., Li, C.H. (1967) Isolation, purification, and characterization of gamma-lipotropic hormone from sheep pituitary glands. *Can J Biochem.* 45(7), pp. 1163–1174.

[7] Seidah, N.G., Gaspar, L., Mion, P., Marcinkiewicz, M., Mbikay, M., Chrétien, M. (1990) cDNA sequence of two distinct pituitary proteins homologous to Kex2 and furin gene products: tissue-specific mRNAs encoding candidates for pro-hormone processing proteinases. *DNA Cell Biol.* 9(6), pp. 415–24. Erratum in *DNA Cell Biol.* 9(10), p. 789.

[8] Smeekens, S.P., Steiner, D.F. (1990) Identification of a human insulinoma cDNA encoding a novel mammalian protein structurally related to the yeast dibasic processing protease Kex2. *J Biol Chem.* 265(6), pp. 2997–3000.

[9] Mbikay, M., Seidah, N.G. (editors) (2011) Proprotein Convertases, *Methods Mol Biol.* Humana press, pp. 1–378.

[10] Seidah, N.G., Benjannet, S., Wickham, L., Marcinkiewicz, J., Bélanger Jasmin, S., Stifani, S., Basak, A., Prat, A., Chrétien M. (2003) The secretory proprotein convertase neural apoptosis-regulated convertase 1 (NARC-1): liver regeneration and neuronal differentiation. *Proc Natl Acad Sci USA.* 100, pp. 928–933.

[11] Chrétien, M., Seidah, N.G., Basak, A., Mbikay, M. (2008) Proprotein convertases as therapeutic targets. *Expert Opin Ther Targets.* 12(10), pp. 1289–1300.

[12] Khatib, A.M., Siegfried, G., Chrétien, M., Metrakos, P., Seidah, N.G. (2002) Proprotein

convertases in tumor progression and malignancy: novel targets in cancer therapy. *Am J Pathol.* 160, pp. 1921–1935.

[13] Scamuffa, N., Calvo, F., Chrétien, M., Seidah, N.G., Khatib, A.M. (2006) Proprotein convertases: lessons from knockouts. *FASEB J.* 20, pp. 1954–1963.

[14] Turpeinen, H., Raitoharju, E., Oksanen, A., Oksala, N., Levula, M., Lyytikäinen, L.P., Järvinen, O., Creemers, J.W., Kähönen, M., Laaksonen, R., Pelto-Huikko, M., Lehtimäki, T., Pesu, M. (2011) Proprotein convertases in human atherosclerotic plaques: the overexpression of furin and its substrate cytokines BAFF and APRIL. *Atherosclerosis.* 219(2), pp. 799–806.

[15] Chitramuthu, B.P., Bennett, H.P. (2011) Use of zebrafish and knockdown technology to define proprotein convertase activity. *Methods Mol Biol.* 768, pp. 273–296.

[16] Page, R.E., Klein-Szanto, A.J., Litwin, S., Nicolas, E., Al-Jumaily, R., Alexander, P., Godwin, A.K., Ross, E.A., Schilder, R.J., Bassi, D.E. (2007) Increased expression of the proprotein convertase furin predicts decreased survival in ovarian cancer. *Cell Oncol.* 29(4), pp. 289–299.

[17] Bassi, D.E., Mahloogi, H., Al-Saleem, L., Lopez De Cicco, R., Ridge, J.A., Klein-Szanto, A.J. (2001) Elevated furin expression in aggressive human head and neck tumors and tumor cell lines. *Mol Carcinog.* 31(4), pp. 224–232.

[18] Cheng, M., Xu, N., Iwasiow, B., Seidah, N., Chrétien, M., Shiu, R.P. (2001) Elevated expression of proprotein convertases alters breast cancer cell growth in response to estrogen and tamoxifen. *J Mol Endocrinol.* 26(2), pp. 95–105.

[19] Basak, A. (2005) Inhibitors of proprotein convertases. *J Mol Med.* 83(11), pp. 844–855.

[20] Kibirev, V.K., Osadchuk, T.V. (2012) Structure and properties of proprotein convertase inhibitors. *Ukr Biokhim Zh.* 84(2), pp. 5–29, (Russian).

[21] Becker, G.L., Lu, Y., Hardes, K., Strehlow, B., Levesque, C., Lindberg, I., Sandvig, K., Bakowsky, U., Day, R., Garten, W., Steinmetzer, T. (2012) Highly potent inhibitors of proprotein convertase furin as potential drugs for treatment of infectious diseases. *J Biol Chem.* 287(26), pp. 21992–22003.

[22] Couture, F., D'Anjou, F., Day, R. (2011) On the cutting edge of proprotein convertase pharmacology: from molecular concepts to clinical applications. *Biomol Concepts.* 2(5), pp. 421–438.

[23] Zhu, J., Declercq, J., Roucourt, B., Ghassabeh, G.H., Meulemans, S., Kinne, J., David, G., Vermonkeu, A.J., Van de ven, W.J., Lindberg, I., Muyldermans, S., Creemers, J.W., (2012) Generation and characterization of non-competitive furin-inhibiting nanobodies. *Biochem J.* Aug 24, Epub, PMID: 22920187.

[24] Kuester, M., Becker, G.L., Hardes, K., Lindberg, I., Steinmetzer, T., Than, M.E. (2011)

Purification of the proprotein convertase furin by affinity chromatography based on PC-specific inhibitors. *J. Biol Chem.* 392(11), pp. 973–981.

[25] Lindberg, I., Appel, J.R. (2011) Inhibitor screening of proprotein convertases using positional scanning libraries. *Methods Mol Biol.* 768, pp. 155–166.

[26] Kibirev, V.K., Osadchuk, T.V., Vadziuk, O.B., Shablykin, O.V., Kozachenko, A.P., Chumachenko, S.A., Popil'nichenko, S.V., Brovarets, V.S. (2011) Study on derivatives of 5-amino-4-acylamino-1H-pyrazole as inhibitors of furin. *Ukr Biokhim Zh.* 83(1), pp. 30–37, (Russian).

[27] Becker, G.L., Hardes, K., Steinmetzer, T. (2011) New substrate analogue furin inhibitors derived from 4-amidinobenzylamide. *Bioorg Med Chem Lett.* 21(16), pp. 4695–4697.

[28] Sielaff, F., Than, M.E., Bevec, D., Lindberg, I., Steinmetzer, T. (2011) New furin inhibitors based on weakly basic amidinohydrazones. *Bioorg Med Chem Lett.* 21(2), pp. 836–840.

[29] Jean, F., Stella, K., Thomas, L., Liu, G., Xiang, Y., Reason, A.J., Thomas, G. (1998) α_1 Antitrypsin Portland, a bioengineered serpin highly selective for furin: application as an antipathogenic agent. *Proc Natl Acad Sci USA.* 95 (13), pp. 7293–7298.

[30] Benjannet, S., Savaria, D., Laslop, A., Munzer, J.S., Chrétien, M., Marcinkiewicz, M., Seidah, N.G. (1997) Alpha1-antitrypsin Portland inhibits processing of precursors mediated by proprotein convertases primarily within the constitutive secretory pathway. *J Biol Chem.* 272(42), pp. 26210–26218.

[31] Bergeron, F., Leduc, R., Day, R. (2000) Subtilase-like proprotein convertases: from molecular specificity to therapeutic applications. *J Mol Endocrinol.* 24, pp. 1–22.

[32] Fugère, M., Day, R. (2005) Cutting back on pro-protein convertases: the latest approaches to pharmacological inhibition. *Trends Pharmacol Sci.* 26, pp. 294–301.

[33] Bergeron, F., Leduc, R., Day, R., et al. (2000) Subtilase-like pro-protein convertases: from molecular specificity to therapeutic applications. *J Mol Endocrinol.* 24, pp. 1–22.

[34] Fugère, M., Limperis, P.C., Beaulieu-Audy, V., Gagnon, F., Lavigne, P., Klarskov, K., Leduc, R., Day, R. (2002) Inhibitory potency and specificity of subtilase-like pro-protein convertase (SPC) prodomains. *J Biol Chem.* 277, pp. 7648–7656.

[35] Artenstein, A.W., Opal, S.M. (2011) Proprotein convertases in health and disease. *N Engl J Med.* 365(26), pp. 2507–2518.

[36] Seidah, N.G., Prat, A. (2012) The biology and therapeutic targeting of the proprotein convertases. *Nat Rev Drug Discov.* 11(5), pp. 367–383.

[37] Thomas, G. (2002) Furin at the cutting edge: from protein traffic to embryogenesis and disease. *Nat Rev Mol Cell Biol.* 3(10), pp. 753–766.

[38] Creemers, J.W., Khatib, A.M. (2008) Knock-out mouse models of pro-protein convertases: unique functions or redundancy? *Front Biosci.* 172, pp. 4960–4971.

[39] Scamuffa, N., Calvo, F., Chrétien, M., Seidah, N.G., Khatib, A-M. (2006) Proprotein convertases: lessons from knockouts. *FASEB J.* 20, pp. 1954–1963.

[40] Seidah, N.G. (2011) What lies ahead for the proprotein convertases? *Ann NY Acad Sci.* 1220, pp. 149–161.

[41] Law, R.H.P., Zhang, Q., McGowan, S., Buckle, A.M., Silverman, G.A., Wong, W., Rosado, C.J., Langendorf, C.G., Pike, R.N., Bird, P.I., Whisstock, J.C. (2006) An overview of the serpin superfamily. *Genome Biol*, 7,216, pp. 1–11.

[42] Richer, M.J., Keays, C.A., Waterhouse, J., Minhas, J., Hashimoto, C., Jean, F. (2004) The Spn4 gene of *Drosophila* encodes a potent furin-directed secretory pathway serpin. *Proc Natl Acad Sci USA.* 101(29), pp. 10560–10565.

[43] Oley, M., Letzel, M.C., Ragg, H. (2004) Inhibition of furin by serpin Spn4A from Drosophila melanogaster. *FEBS Lett.* 577(1–2), pp. 165–169.

[44] Bentele, C., Krüge, O., Tödtmann, U., Oley, M., Ragg, H.A. (2006) Proprotein convertase-inhibiting serpin with an endoplasmic reticulum targeting signal from *Branchiostoma lanceolatum*, a close relative of vertebrates. *Biochem J.* 395(3), pp. 449–456.

[45] Leblond, J., Laprise, M.H., Gaudreau, S., Grondin, F., Kisiel, W., Dubois, C.M. (2006) The serpin proteinase inhibitor 8: an endogenous furin inhibitor released from human platelets. *Thromb Haemost.* 95(2), pp. 243–252.

[46] Bernot, D., Stalin, J., Stocker, P., Bonardo, B., Scroyen, I., Alessi, M.C., Peiretti, F. (2011) Plasminogen activator inhibitor 1 is an intracellular inhibitor of furin proprotein convertase. *J Cell Sci.* 124(Pt 8), pp. 1224–1230.

[47] Cornwall, G.A., Cameron, A., Lindberg, I., Hardy, D.M., Cormier, N., Hsia, N. (2003) The cystatin-related epididymal spermatogenic protein inhibits the serine protease prohormone convertase 2. *Endocrinology.* 144(3), pp. 901–908.

[48] Börner, S., Ragg, H. (2011) Bioinformatic approaches for the identification of serpin genes with multiple reactive site loop coding exons. *Methods Enzymol.* 501, pp. 209–222.

[49] Mbikay, M., Seidah, N.G., Chrétien, M. (2001) Neuroendocrine secretory protein 7B2: structure, expression and functions. *Biochem J.* 357(Pt 2), pp. 329–342.

[50] Fricker, L.D., McKinzie, A.A., Sun, J., Curran, E., Qian, Y., Yan, L., Patterson, S.D., Courchesne, P.L., Richards, B., Levin, N., Mzhavia, N., Devi, L.A., Douglass, J. (2000) Identification and characterization of proSAAS, a granin-like neuroendocrine peptide precursor that inhibits prohormone processing. *J Neurosci.* 20(2), pp. 639–648.

[51] Jean, F., Stella, K., Thomas, L., Liu, G., Xiang, Y., Reason, A.J., Thomas, G. (1998) alpha1-Antitrypsin Portland, a bioengineered serpin highly selective for furin: application as an antipathogenic agent. *Proc Natl Acad Sci USA.* 95(13), pp. 7293–7298.

[52] Dufour, E.K., Désilets, A., Longpré, J.M., Leduc, R. (2005) Stability of mutant serpin/furin

complexes: dependence on pH and regulation at the deacylation step *Protein Sci.* 14(2), pp. 303–315.

[53] Mercapide, J., Lopez De Cicco, R., Bassi, D.E., Castresana, J.S., Thomas, G., Klein-Szanto, A.J. (2002) Inhibition of furin-mediated processing results in suppression of astrocytoma cell growth and invasiveness. *Clin Cancer Res.* 8(6), pp. 1740–1746.

[54] Chen, R.N., Huang, Y.-H., Lin, Y.C., Yeh, C.-T., Liang, Y., Chen, S.-L. and Lin, K.-H. (2008) Thyroid hormone promotes cell invasion through activation of furin expression in human hepatoma cell lines. *Endocrinology.* 149(8), pp. 3817–3831

[55] http://www.sumobrain.com/patents/wipo/Furin-biologically-active-derivatives-thereof/WO2011144517.html; WIPO Patent Application WO/2006/084131, inventors: Cohen, S.M., Khatib, A.-M.

[56] Kayo, T., Sawada, Y., Suda, M., Konda, Y., Izumi, T., Tanaka, S., Shibata, H., Takeuchi, T. (1997) Proprotein-processing endoprotease furin controls growth of pancreatic beta-cells. *Diabetes.* 46(8), pp. 1296–1304.

[57] Benjannet, S., Savaria, D., Laslop, A., Munzer, J.S., Chrétien, M., Marcinkiewicz, M., Seidah, N.G. α1-Antitrypsin Portland inhibits processing of precursors mediated by proprotein convertases primarily within the constitutive secretory pathway. *J Biol Chem.* 272, pp. 26210–26218.

[58] Lu, W., Zhang, W., Molloy, S.S., Thomas, G., Ryan, K., Chiang, Y., Anderson, S., Laskowski, M. Jr (1993) Arg15–Lys17–Arg18 turkey ovomucoid third domain inhibits human furin. *J Biol Chem.* 268, pp. 14583–14585.

[59] Van Rompaey, L., Proost, P., Van den Berghe, H., Marynen, P. (1995) Design of a new protease inhibitor by the manipulation of the bait region of alpha 2-macroglobulin: inhibition of the tobacco etch virus protease by mutant alpha 2-macroglobulin. *Biochem J.* 312(Pt 1), pp. 191–195.

[60] Komiyama, T., VanderLugt, B., Fugère, M., Day, R., Kaufman, J.R., Fuller, R.S. (2003) Optimization of protease-inhibitor interactions by randomizing adventitious contacts. *Proc Natl Acad Sci USA.* 100(14), pp. 8205–8210.

[61] Basak, A., Koch, P., Dupelle, M., Fricker, L.D., Devi, L.A., Chrétien, M., Seidah, N.G. (2001) Inhibitory specificity and potency of proSAAS-derived peptides toward proprotein convertase 1. *J Biol Chem.* 276(35), pp. 32720–32728.

[62] Sarac, M.S., Peinado, J.R., Leppla, S.H., Lindberg, I. (2004) Protection against anthrax toxemia by hexa-D-arginine in vitro and in vivo. *Infect Immun.* 72(1), pp. 602–605.

[63] Sarac, M.S., Cameron, A., Lindberg, I. (2002) The furin inhibitor hexa-D-arginine blocks the activation of *Pseudomonas aeruginosa* exotoxin A in vivo. *Infect Immun.* 70(12), pp. 7136–7139.

[64] Cameron, A., Appel, J., Houghten, R.A., Lindberg, I. (2000) Polyarginines are potent furin inhibitors. *J Biol Chem.* 275(47), pp. 36741–36749.

[65] Basak, A., Chen, A., Scamuffa, N., Mohottalage, D., Basak, S., Khatib, A.M. (2010) Blockade of furin activity and furin-induced tumor cells malignant phenotypes by the chemically synthesized human furin prodomain. *Curr Med Chem.* 17(21), pp. 2214–2221.

[66] Majumdar, S., Chen, A., Palmer-Smith, H., Basak, A. (2011) Novel circular, cyclic and acyclic Ψ(CH(2)O) containing peptide inhibitors of SKI-1/S1P: synthesis, kinetic and biochemical evaluations. *Curr Med Chem.* 18(18), pp. 2770–2782.

[67] Jean, F., Basak, A., DiMaio, J., Seidah, N.G., Lazure, C. (1995) An internally quenched fluorogenic substrate of prohormone convertase 1 and furin leads to a potent prohormone convertase inhibitor. *Biochem J.* 307(Pt 3), pp. 689–695.

[68] Basak, A., Khatib, A.M., Mohottalage, D., Basak, S., Kolajova, M., Bag, S.S., Basak, A. (2009) A novel enediynyl peptide inhibitor of furin that blocks processing of proPDGF-A, B and proVEGF-C. *PLoS One.* 26; 4(11), pp. e7700.

[69] Becker, G.L., Sielaff, F., Than, M.E., Lindberg, I., Routhier, S., Day, R., Lu, Y., Garten, W., Steinmetzer, T. (2010) Potent inhibitors of furin and furin-like proprotein convertases containing decarboxylated P1 arginine mimetics. *J Med Chem.* 53(3), pp. 1067–1075.

[70] Smith, H.J., Simons, C. (2002) Proteinase and peptidase inhibition: recent potential targets for drug development (Google eBook), CRC Press.

[71] Cragg, G.M., Newman, D.J. (2009) Drug Discovery and development from natural products: The way forward. 11th NAPRECA Symposium Book of Proceedings, Antananarivo, Madagascar, pp. 56–69.

[72] Cragg, G.M., Grothaus, P.G., Newman, D.J. (Editor Valdir Cechinel-Filho), (2009), Plant bioactives and drug discovery, Wiley & Sons Inc, New Hoboken, New Jersey, USA, pp. 1–27

[73] Lipinski, C.A., Lombardo, F., Dominy, B.W., Feeney, P.J. (1997) Experimental and computational approaches to estimate solubility and permeability in drug discovery and development settings. *Adv Drug Delivery Rev.* 23, pp. 3–25.

[74] Lipinski, C.A. (2000) Drug-like properties and the causes of poor solubility and poor permeability. *J Pharm Tox Meth.* 44, pp. 235–249.

[75] Basak, A., Cooper, S., Roberge, A.G., Banik, U.K., Chrétien, M., Seidah, N.G. (1999) Inhibition of proprotein convertases, furin, PC1 and PC7 by diterpines of *Andrographis paniculata* and their succinoyl esters. *Biochem J.* 338 (1), pp. 107–113.

[76] Jiao, G.S., Cregar, L., Wan, J., Millis, S.Z., Tang, C., O'Malley, S., Johnson, A.T., Sareth, S., Larson, J., Thomas, G. (2006) Synthetic small molecule furin inhibitors derived from 2,5-dideoxystreptamine. *Proc Natl Acad Sci USA.* 103(52), pp. 19707–19712.

[77] Jiao, G.S., Cregar, L., Goldman, M.E., Millis, S.Z., Tang, C. (2006) Guanidinylated 2,

5-dideoxystreptamine derivatives as anthrax lethal factor inhibitors. *Bioorg Med Chem Lett.* 16(6), pp. 1527–1531.

[78] Vivoli, M., Caulfield, T.R., Martínez-Mayorga, K., Johnson, A.T., Jiao, G.S., Lindberg, I. (2012) Inhibition of prohormone convertases PC1/3 and PC2 by 2, 5-dideoxystreptamine derivatives. *Mol Pharmacol.* 81(3), pp. 440–454.

[79] Kowalska, D., Liu, J., Appel, J.R., Ozawa, A., Nefzi, A., Mackin, R.B., Houghten, R.A., Lindberg, I. (2009) Synthetic small-molecule prohormone convertase 2 inhibitors. *Mol Pharmacol.* 75(3), pp. 617–625.

[80] Hook, V., Funkelstein, L., Toneff, T., Mosier, C., Hwang, S.R. (2009) Human pituitary contains dual cathepsin L and prohormone convertase processing pathway components involved in converting POMC into the peptide hormones ACTH, alpha-MSH, and beta-endorphin. *Endocrine.* 35(3), pp. 429–437.

[81] Podsiadlo, P., Komiyama, T., Fuller, R.S., Blum, O. (2004) Furin inhibition by compounds of copper and zinc. *J Biol Chem.* 279(35), pp. 36219–36227.

[82] Basak, A., Mandal, S., Bag, S.S. (2003) Chelation-controlled Bergman cyclization: Synthesis and reactivity of enediynyl ligands. *Chem. Rev.* 103, 4077–4094.

[83] Majumdar, S., Mohanta, B.C., Roy Chowdhury, D., Banik, R., Dinda, B.N., Basak, A. (2010) Proprotein convertase inhibitory activities of flavonoids isolated from *Oroxylum indicum*. *Curr Med Chem.* 17 (19), pp. 2049–2058.

[84] Kibirev, V.K., Osadchuk T.V., Vadziuk, O.B., Garazd, M.M. (2010) New non-peptide inhibitors of furin. *Ukr Biokhim Zh.* 82(2), pp. 15–21 (Russian).

[85] Komiyama, T., Coppola, J.M., Larsen, M.J., van Dort, M.E., Ross, B.D., Day, R., Rehemtulla, A., Fuller, R.S. (2009) Inhibition of furin/proprotein convertase-catalyzed surface and intracellular processing by small molecules. *J Biol Chem.* 284(23), pp. 15729–15738.

[86] López-Vallejo, F., Martínez-Mayorga, K. (2012) Furin inhibitors: importance of the positive formal charge and beyond. *Bioorg Med Chem.* 20 (14), pp. 4462–4471.

[87] Cheng, Y., Prusoff, W. (1973) Relationship between the inhibition constant (Ki) and the concentration of inhibitor which causes 50 per cent inhibition (IC_{50}) of an enzymatic reaction. *Biochem Pharmacol.* 22(23), pp. 3099–3108.

[88] Dinda, B., Debnath, S., Banik, R. (2011) Naturally occurring iridoids and secoiridoids. An updated review, part 4. *Chem Pharm Bull (Tokyo).* 59(7), pp. 803–833.

[89] Dinda, B., Chowdhury, D.R., Mohanta, B.C. (2009) Naturally occurring iridoids, secoiridoids and their bioactivity. An updated review, part 3. *Chem Pharm Bull (Tokyo).* 57(8), pp. 765–796.

[90] De, U.C., Ghosh, R. Chowdhury, S., Dinda, B. (2012) New iridoid from aerial parts of *Mussaenda roxburghii*. *Natural Product Commun.* 7(1), pp. 1–2.

[91] Becker, G.L., Lu, Y., Hardes, K., Strehlow, B., Levesque, C., Lindberg, I., Sandvig, K.,

Bakowsky, U., Day, R., Garten, W., Steinmetzer, T. (2012) Highly potent inhibitors of proprotein convertase furin as potential drugs for treatment of infectious diseases. *J Biol Chem.* 287(26), pp. 21992–22003.

[92] Becker, G.L., Hardes, K., Steinmetzer, T. (2011) New substrate analogue furin inhibitors derived from 4-amidinobenzylamide. *Bioorg Med Chem Lett.* 21(16), pp. 4695–4697.

[93] Coppola, J.M., Hamilton, C.A., Bhojani, M.S., Larsen, M.J., Ross, B.D., Rehemtulla, A. (2007) Identification of inhibitors using a cell based assay for monitoring golgi-resident protease activity. *Anal Biochem.* 364(1), pp. 19–29.

[94] Hamstra, D.A., Rehemtulla, A. (1999) Towards an enzyme/prodrug strategy for cancer gene therapy: endogenous activation of carboxypeptidase A mutants by the PACE/Furin family of propeptidases. *Hum Gene Ther.* 10, pp. 235–248.

[95] Basak, A., Mitra, D., Mohottalage, D., Basak, A. (2010) C2-Symmetric azobenzene-amino acid conjugates and their inhibitory properties against Subtilisin Kexin Isozyme-1. Submitted to *Bioorg Med Chem Lett.* 20(13), pp. 3977–3981.

[96] Khatib, Abdel-Majid (Ed.) (2006) Regulation of carcinogenesis, angiogenesis and metastasis by the proprotein convertases (PC's): A new potential strategy in cancer therapy, pp. 1–158, Publisher: Springer

[97] Bahbouhi, B., Bendjennat, M., Guétard, D., Seidah, N.G., Bahraoui, E. (2000) Effect of alpha-1 antitrypsin Portland variant (alpha 1-PDX) on HIV-1 replication. *Biochem J.* 352(Pt 1), pp. 91–98.

[98] Benjannet, S., Savaria, D., Laslop, A., Munzer, J.S., Chrétien, M., Marcinkiewicz, M., Seidah, N.G. (1997) Alpha1-antitrypsin Portland inhibits processing of precursors mediated by proprotein convertases primarily within the constitutive secretory pathway. *J Biol Chem.* 272(42), pp. 26210–26218.

[99] Tsuji, A., Hashimoto, E., Ikoma, T., Taniguchi, T., Mori, K., Nagahama, M., Matsuda, Y. (1999) Inactivation of proprotein convertase, PACE4, by alpha1-antitrypsin Portland (alpha1-PDX), a blocker of proteolytic activation of bone morphogenetic protein during embryogenesis: evidence that PACE4 is able to form an SDS-stable acyl intermediate with alpha1-PDX. *J Biochem.* 126(3), pp. 591–603.

[100] Bassi, D.E., De Cicco, R.L., Mahloogi, H., Zucker, S., Thomas, G., Klein-Szanto, A.J.P. (2001) Furin inhibition results in absent or decreased invasiveness and tumorigenicity of human cancer cells, *Proc Natl Acad Sci USA.* 98(18), pp. 10326–10331.

[101] Kines, R.C., Thompson, C.D., Lowy, D.R., Schiller, J.T., Day, P.M. (2009) The initial steps leading to papillomavirus infection occur on the basement membrane prior to cell surface binding. *Proc Natl Acad Sci USA.* 106 (48), pp. 20458–20463.

[102] http://www.merckmillipore.com/is-bin/INTERSHOP.enfinity.

[103] Winston, A., Back, D., Fletcher, C., et al. (2006) Effect of omeprazole on the pharmacokinetics of saquinavir-500 mg formulation with ritonavir in healthy male and female volunteers. *AIDS.* 20 (10), pp. 401–406.

[104] Shen, C.H., Wang, Y.F., Kovalevsky, A.Y., Harrison, R.W., Weber, I.T. (2010) Amprenavir complexes with HIV-1 protease and its drug-resistant mutants altering hydrophobic clusters. *FEBS J.* 277(18), pp. 3699–3714.

[105] Ghosh, A.K., Dawson, Z.L., Mitsuya, H. (2007) Darunavir, a conceptually new HIV-1 protease inhibitor for the treatment of drug-resistant HIV. *Bioorg Med Chem.* 15(24), pp. 7576–7580.

[106] KALETRA (lopinavir/ritonavir) capsules, (lopinavir/ritonavir) oral solution. Prescribing information. April 2009, www.rxlist.com/kaletra-capsules-drug.htm.

[107] Capparelli, E., Holland, D., Okamoto, C., et al. (2005) Lopinavir concentrations in cerebrospinal fluid exceed the 50% inhibitory concentration for HIV. *AIDS.* 19(9), pp. 949–952.

[108] Liu, F., Boross, P.I., Wang, Y.F., Tozser, J., Louis, J.M., Harrison, R.W., Weber, I.T. (2005) Kinetic, stability, and structural changes in high-resolution crystal structures of HIV-1 protease with drug-resistant mutations L24I, I50V, and G73S. *J Mol Biol.* 354(4), pp. 789–800.

[109] Eira, M., Araujo, M., Seguro, A.C. (2006) Urinary NO3 excretion and renal failure in indinavir-treated patients. *Braz J Med Biol Res.* 39, pp. 1065–1070.

[110] Shankar, S.S., Dubé, M.P., Gorski, J.C., Klaunig, J.E., Steinberg, H.O. (2005) Indinavir impairs endothelial function in healthy HIV-negative men. *Am Heart J.* 150(5), pp. 933.e1–933.e7.

[111] Bauer, J., et al. (2004) Ritonavir: an extraordinary example of conformational polymorphism. *Pharm Res.* 18(6), pp. 859–866.

[112] Zeldin, R.K., Petruschke, R.A. (2004) Pharmacological and therapeutic properties of ritonavir-boosted protease inhibitor therapy in HIV-infected patients. *J Antimicrob Chemother.* 53(1), pp. 4–9.

[113] Eron, J. Jr, Yeni, P., Gathe, J. Jr, et al. (2006) The KLEAN study of fosamprenavir–ritonavir versus lopinavir–ritonavir, each in combination with abacavir–lamivudine, for initial treatment of HIV infection over 48 weeks: a randomised non-inferiority trial. *Lancet.* 368(9534), pp. 476–482.

[114] FDA approval of Isentress (raltegravir). U.S. Food and Drug Administration. June 25, 2009, www.fda.gov/ForConsumers/ByAudience/ForPatientAdvocates/HIVandAIDSActivities/ucm124040.htm. Retrieved 2009-11-15.

[115] Bristol wins U.S. approval for single anti-HIV pill. Reuters. 2006-10-20. http://today.reuters.co.uk/news/articlenews.aspx?type=healthNews&storyID=2006-10-20T184537Z_

01_WEN7494_RTRIDST_0_HEALTH-BRISTOLMYERS-DC.XML&WTmodLoc=SciHealth-C4-Health-5. Retrieved 2006-10-25.

[116] Bardsley-Elliot, A., Plosker, G.L. (2000) Nelfinavir: an update on its use in HIV infection. *Drugs.* 59(3), pp. 581–620.

[117] Njoroge, F.G., Chen, K.X., Shih, N.Y., Piwinski, J.J. (2008) Challenges in modern drug discovery: a case study of boceprevir, an HCV protease inhibitor for the treatment of hepatitis C virus infection. *Acc Chem Res.* 41(1), pp. 50–59.

[118] Yamashita, M., Tomozawa, T., Kakuta, M., Tokumitsu, A., Nasu, H., Kubo, S. (2009) CS-8958, a prodrug of the new neuraminidase inhibitor R-125489, shows long-acting anti-influenza virus activity. *Antimicrob Agents Chemother.* 53(1), pp. 186–192.

[119] Hayden, F. (2009) Developing new antiviral agents for influenza treatment: what does the future hold? *Clin Infect Dis.* 48(Suppl 1), pp. S3–13.

[120] Lew, W., Chen, X., Kim, C.U. (2000) Discovery and development of GS 4104 (oseltamivir): an orally active influenza neuraminidase inhibitor. *Curr Med Chem.* 7(6), pp. 663–672.

[121] Revill, P., Serradell, N., Bolos, J., Rosa, E. (2007) "Telaprevir." *Drugs Future.* 32(9), pp. 788–798.

[122] Lin, C., Kwong, A.D., Perni, R.B. (2006) Discovery and development of VX-950, a novel, covalent, and reversible inhibitor of hepatitis C virus NS3.4A serine protease. *Infect Disord Drug Targets.* 6(1), pp. 3–16.

[123] www.cdc.gov/flu/weekly/weeklyarchives 2008-2009/weekly32.htm.

[124] von Itzstein, M., Wu, W.Y., Kok, G.B., et al. (1993) Rational design of potent sialidase-based inhibitors of influenza virus replication. *Nature.* 363(6428), pp. 418–423.

[125] Peták, I., Mihalik, R., Bauer, P.I., Süli-Vargha, H., Sebestyén, A., Kopper, L. (1998) BCNU is a caspase-mediated inhibitor of drug-induced apoptosis. *Cancer Res.* 58(4), pp. 614–618.

[126] Epivir package insert (PDF). Glaxo Smith Kline. http://us.gsk.com/products/assets/us_epivir.pdf.

[127] Fox, Z., Dragsted, U.B., Gerstoft, J. et al. (2006) A randomized trial to evaluate continuation versus discontinuation of lamivudine in individuals failing a lamivudine-containing regimen: the COLATE trial. *Antivir Ther.* 11(6), pp. 761–770.

[128] Mallal, S., Nolan, D., Witt, C. et al. (2002) Association between the presence of HLA-B 5701, HLA-DR7 and HLA-DQ3 and hypersensitivity to HIV-1 reverse transcriptase inhibitor abacavir. *Lancet.* 359(9308), pp. 727–732.

[129] Wright, K. (1986) AIDS therapy. First tentative signs of therapeutic promise. *Nature.* 323(6086), p. 283

[130] WHO Model List of Essential Medicines (PDF) World Health Organization. (2005) http://whqlibdoc.who.int/hq/2005/a87017_eng.pdf. Retrieved 2006-03-12.

[131] Doyon, L., Tremblay, S., Bourgon, L., Wardrop, E., Cordingley, M. (2005) Selection and characterization of HIV-1 showing reduced susceptibility to the non-peptidic protease inhibitor tipranavir. *Antivir Res.* 68(1), pp. 27–35.

[132] New Aptivus (tipranavir) Oral Solution Approved for Treatment-Experienced Pediatric and Adolescent HIV Patients (Press release). *Boehringer Ingelheim*. 2008-06-24.

TITLES OF RELATED INTEREST

Colloquium Series on Neuropeptides

Editors

Lakshmi Devi, Ph.D., *Professor, Department of Pharmacology and Systems Therapeutics, Associate Dean for Academic Enhancement and Mentoring, Mount Sinai School of Medicine, New York*

Lloyd D. Fricker, Ph.D., *Professor, Department of Molecular Pharmacology, Department of Neuroscience, Albert Einstein College of Medicine, New York*

Communication between cells is essential in all multicellular organisms, and even in many unicellular organisms. A variety of molecules are used for cell-cell signaling, including small molecules, proteins, and peptides. The term 'neuropeptide' refers specifically to peptides that function as neurotransmitters, and includes some peptides that also function in the endocrine system as peptide hormones. Neuropeptides represent the largest group of neurotransmitters, with hundreds of biologically active peptides and dozens of neuropeptide receptors known in mammalian systems, and many more peptides and receptors identified in invertebrate systems. In addition, a large number of peptides have been identified but not yet characterized in terms of function. The known functions of neuropeptides include a variety of physiological and behavioral processes such as feeding and body weight regulation, reproduction, anxiety, depression, pain, reward pathways, social behavior, and memory. This series will present the various neuropeptide systems and other aspects of neuropeptides (such as peptide biosynthesis), with individual volumes contributed by experts in the field.

For a list of published and forthcoming titles:
http://www.morganclaypool.com/toc/npe/1/1

Colloquium Series on The Developing Brain

Editor

Margaret McCarthy, PhD., *Professor, Department of Physiology; Associate Dean for Graduate Studies; and, Acting Chair, Department of Pharmacology & Experimental Therapeutics, University of Maryland School of Medicine*

The goal of this series is to provide a comprehensive state-of-the art overview of how the brain develops and those processes that affect it. Topics range from the fundamentals of axonal guidance and synaptogenesis prenatally to the influence of hormones, sex, stress, maternal care and injury during the early postnatal period to an additional critical period at puberty. Easily accessible expert reviews combine analyses of detailed cellular mechanisms with interpretations of significance and broader impact of the topic area on the field of neuroscience and the understanding of brain and behavior.

For a list of published and forthcoming titles:
http://www.morganclaypool.com/toc/dbr/1/1

Colloquium Series on The Genetic Basis of Human Disease

Editor

Michael Dean, Ph.D., *Head, Human Genetics Section, Senior Investigator, Laboratory of Experimental Immunology National Cancer Institute (at Frederick)*

This series will explore the genetic basis of human disease, documenting the molecular basis for rare and common; Mendelian and complex conditions. The series will overview the fundamental principles in understanding such as Mendel's laws of inheritance, and genetic mapping through modern examples. In addition current methods (GWAS, genome sequencing) and hot topics (epigenetics, imprinting) will be introduced through examples of specific diseases.

For a full list of published and forthcoming titles:
http://www.morganclaypool.com/page/gbhd